The Zynq® Book Tutorials

for Zybo and ZedBoard

The Zynq® Book Tutorials

for Zybo and ZedBoard

Louise H. Crockett

Ross A. Elliot

Martin A. Enderwitz

David Northcote

Series Editors: Louise H. Crockett and Robert W. Stewart

Department of Electronic and Electrical Engineering

University of Strathclyde

Glasgow, Scotland, UK

August 2015

This edition first published August 2015 by Strathclyde Academic Media.

Open Source Licence to Use and Reproduce

Tutorial Files

Tutorial files are distributed via the book's companion website: www.zynqbook.com.

Warning and Disclaimer

Trademarks

Acknowledgements

This is a new version of the tutorials accompanying The Zynq Book. It is based in Vivado 2015.1, and now supports both the ZedBoard and the Zybo development boards.

A number of people contributed valuable of feedback on the original set of tutorials, on which these are based. Austin Lesea and Y. C. Wang at Xilinx tested the tutorials at an early stage in their creation, and gave us several useful suggestions. At the University of Strathclyde, Iain Chalmers, Sarunas Kalade, Damien Muir and Craig Ramsay have also been also greatly helpful in working through various versions of the tutorials and telling us about their user experiences.

Once again, our sincerest thanks must go to Cathal McCabe of Xilinx University Program, who has not only provided vital feedback and support in the creation of the tutorial material, but has also coordinated the distribution of those materials to others.

Louise Crockett, Ross Elliot, Martin Enderwitz, and David Northcote.

August 2015.

How to Use This Book

Example Files and Ebook Version

In order to follow The Zynq Book Tutorials, you should download a set of prepared files from the book's website:

www.zynqbook.com

An electronic book (non-printable PDF) version of this set of tutorials can also be downloaded from the above link.

Instructions for Zybo and ZedBoard Development Boards

As you read through the tutorials, you will notice that certain procedures have different variations depending on the development board being used. Where a sequence of instructions is board-specific (i.e. relating either to the ZedBoard or the Zybo), the start of the sequence is indicated by a coloured block icon in the left hand margin:

Zed for ZedBoard

Zybo for Zybo

The resumption of instructions common to both boards is marked with another icon:

Resume

Simply pick out the instructions relevant to your board, by identifying either the Zed or Zybo icon, and then look forward to find where to Resume the main flow.

Operating System

The Zynq Book Tutorials have been tested using the Microsoft Windows operating system. It is expected that they will also function on the Linux Kernel OS, although this has not been tested.

Contents

The Zynq Book Tutorials 1

First Designs on Zynq

v1.5, June 2015

Revision History

Date	Version	Changes
14/06/2013	1.0	First release for Vivado Design Suite version 2013.1
19/06/2013	1.1	Updated for changes in Vivado Design Suite version 2013.2
27/01/2014	1.2	Updated for changes in Vivado Design Suite version 2013.4
30/04/2014	1.3	Updated for changes in Vivado Design Suite version 2014.1
1/04/2015	1.4	Updated for changes in Vivado Design Suite version 2014.4
13/04/2015	1.4.1	Updated to include Zybo development board for Vivado Design Suite version 2014.4
18/06/2015	1.5	Updated for changes in Vivado Design Suite Version 2015.1

Introduction

This tutorial will guide you through the process of creating a first Zynq design using the Vivado™ Integrated Development Environment (IDE), and introduce the IP Integrator environment for the generation of a simple Zynq processor design to be implemented on a Zynq development board. The Software Development Kit (SDK) will then be used to create a simple software application which will run on the Zynq's ARM Processing System (PS) to control the hardware that is implemented in the Programmable Logic (PL).

The tutorial is split into three exercises, and is organised as follows:

Exercise 1A - This exercise will guide you through the process of launching Vivado IDE and creating a project for the first time. The various stages of the *New Project Wizard* will be introduced.

Exercise 1B - In this exercise, we will use the project that was created in Exercise 1A to build a simple Zynq embedded system with the graphical tool, IP Integrator, and incorporating existing IP from the Vivado IP Catalog. A number of design aids will be used throughout this exercise, such as the Board Automation feature which automates the customisation of IP modules for a specified device or board. The Designer Assistance feature, which assists with the connections between the Zynq PS and the IP modules in the PL will also be demonstrated.

Once the design is finished, a number of stages will be undertaken to complete the hardware system and generate a bitstream for implementation in the PL. The completed hardware design will then be exported to the Software Development Kit (SDK) for the development of a simple software application in **Exercise 1C.**

Exercise 1C - In this short third exercise, the SDK will be introduced, and a simple software application will be created to allow the Zynq processor to interact with the IP implemented in the PL. A connection to the hardware server that allows the SDK to communicate with the Zynq processors will be established. The software drivers that are automatically created by the Vivado IDE for IP modules will be explored and integrated into the software application, before finally building and executing the software application on the Zynq.

NOTE: Throughout all of the practical tutorial exercise we will be using **C:\Zynq_Book** as the working directory. If this is not suitable, you can substitute it for a directory of your choice, but you should be aware that you will be required to make alterations to some source files in order to complete exercises successfully.

Exercise 1A Creating a First IP Integrator Design

In this exercise we will create a new project in Vivado IDE by moving through the stages of the Vivado IDE *New Project Wizard.*

Zybo The Zybo requires a **one time** additional set-up procedure in order to set the *Default Part* correctly. This is necessary as Vivado 2015.1 does not contain a board part for the Zybo development board. If you have already configured Vivado 2015.1 with the Zybo board part, you can skip this procedure and start from Step (a).

Open windows explorer and navigate to the following location within the Zynq book source files:

C:\Zynq_Book\sources\zybo\setup\board_part

In this directory you will see a file named ***zybo***. This contains the board part for the **Rev. B.3** Zybo development board. You may also check the revision of your Zybo by inspecting the bottom side of your board. Updated board parts can be retrieved from the Digilent Website using the following link:

https://reference.digilentinc.com/vivado:boardfiles

Copy the ***zybo*** file by right clicking on the file and selecting ***copy*** as shown below:

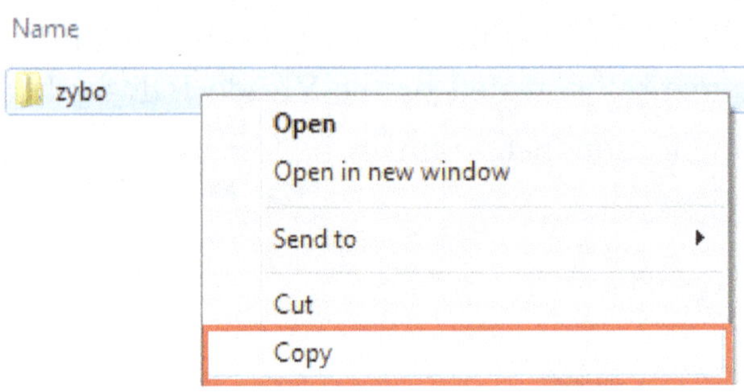

Open a second windows explorer and navigate to the following location in the Vivado 2015.1 installation directory:

{Vivado installation directory}\2015.1\data\boards\board_parts\zynq

This directory is responsible for all the board parts of different Zynq boards that can be used in the Vivado 2015.1 design suite. We will now be adding the Zybo development board to the directory. You may find that a file named ***zybo*** already exists, ignore this and carry on with the following procedure.

Right click on a blank space in the folder and select ***paste*** as shown below:

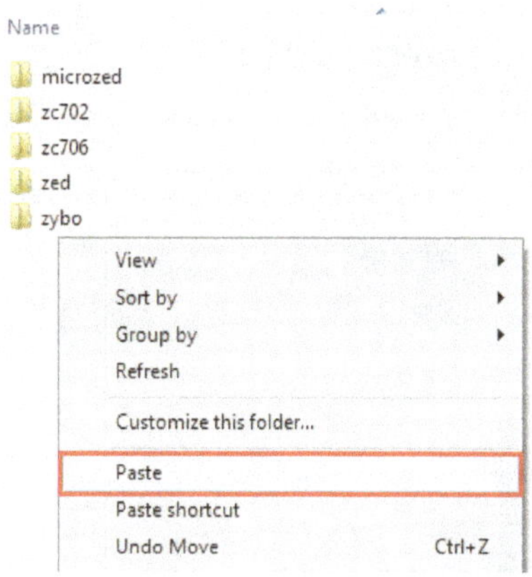

A dialogue window may appear asking to *merge* the incoming folder if a ***zybo*** folder currently exists. Click ***Yes***.

You have now successfully added the Zybo board part to the Vivado 2015.1 Design Suite.

Resume We will start by launching the Vivado IDE.

(a) Launch Vivado by double-clicking on the Vivado desktop icon: , or by navigating to **Start > All Programs > Xilinx Design Tools > Vivado 2015.1> Vivado 2015.1**

(b) When Vivado loads, you will be presented with the *Getting Started* screen as in Figure 1.1.

Figure 1.1: Vivado IDE Getting Started Screen

(c) Select the option to ***Create New Project*** and the *New Project Wizard* will open, as in Figure 1.2.

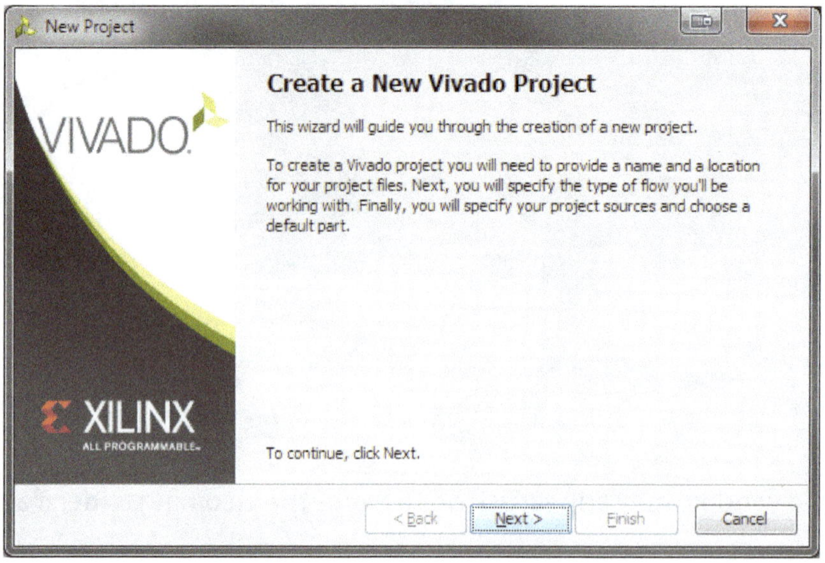

Figure 1.2: New Project Dialogue

Click ***Next***.

(d) At the Project Name dialogue, enter ***first_zynq_design*** as the ***Project name*** and ***C:/Zynq_Book*** as ***Project location***.

Make sure that you select the option to ***Create project subdirectory***. All options should be the same as shown below:

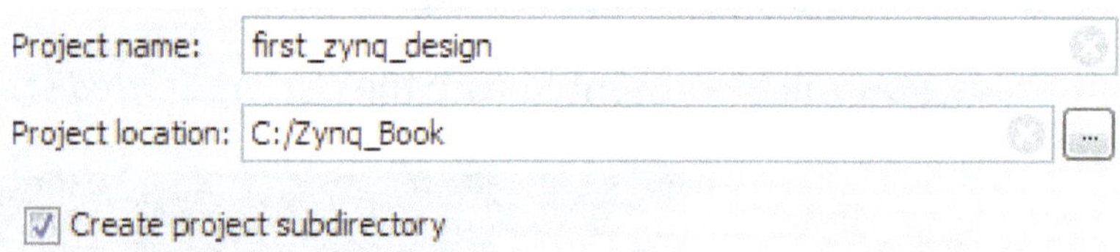

Click ***Next***.

A directory named ***Zynq_Book*** will be created on your ***C drive*** if it did not already exist.

(e) At the *Project Type* dialogue, select ***RTL Project*** and ensure that the option ***Do not specify sources at this time*** is <u>not</u> selected:

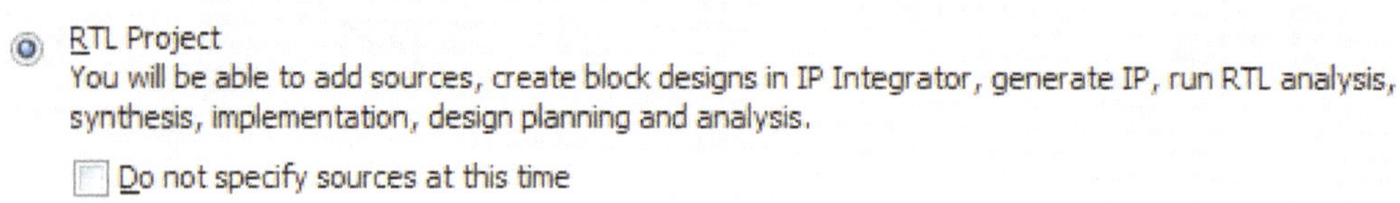

Click ***Next***.

(f) Select ***VHDL*** as the ***Target language*** and ***Mixed*** as the ***Simulator Language*** in the *Add Sources* dialogue:

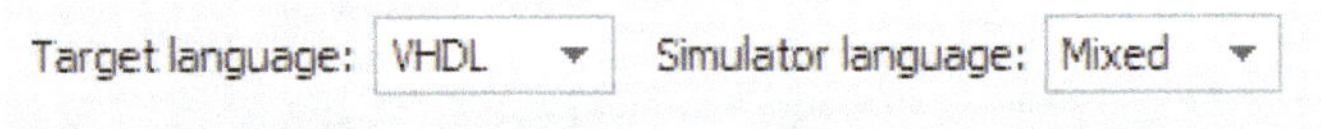

If existing sources, in the form of HDL or netlist files, were to be added to the project they could be imported at this stage.

As we do not have any sources to add to the project, click ***Next***.

(g) The *Add Existing IP (optional)* dialogue will open.

If existing IP sources were to be included in the project, they could be added here.

As we do not have any existing IP to add, click ***Next***.

(h) The *Add Constraints (optional)* dialogue will open.

This is the stage where any physical or timing constraints files could be added to the project.

As we do not have any constraints files to add, click ***Next***.

(i) From the *Default Part* dialogue window,

Zed

Select ***Boards*** from the *Select* dialogue, click ***ZedBoard Zynq Evaluation and Development Kit*** from the *Display Name* list and **All** from the *Board Rev* list, as shown in Figure 1.3. Select the appropriate revision for your board (in this case **Rev. D** has been selected).

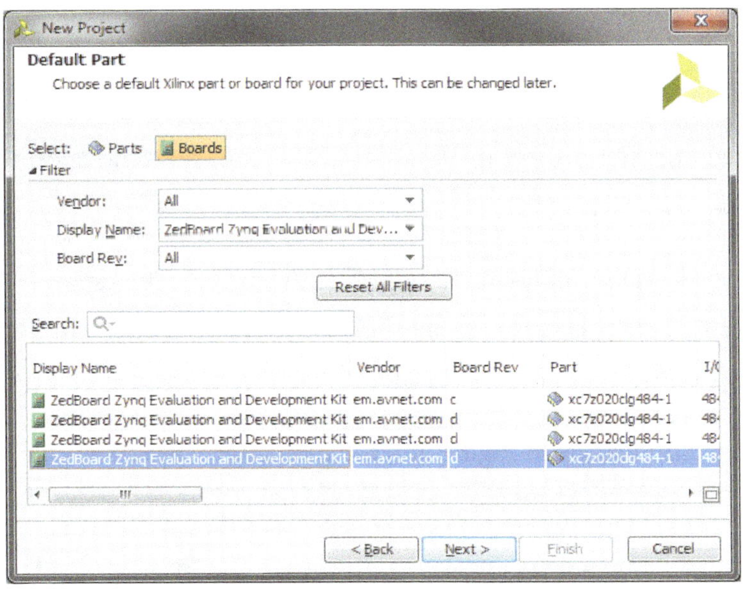

Figure 1.3: Zedboard Default Part Dialogue Options

Click ***Next***.

Zybo

Select ***Boards*** from the *Select* dialogue, click ***Zybo*** from the *Display Name* list and **All** from the *Board Rev* list, as shown in Figure 1.4. Select the appropriate revision for your board (in this case **Rev. B.3** has been selected).

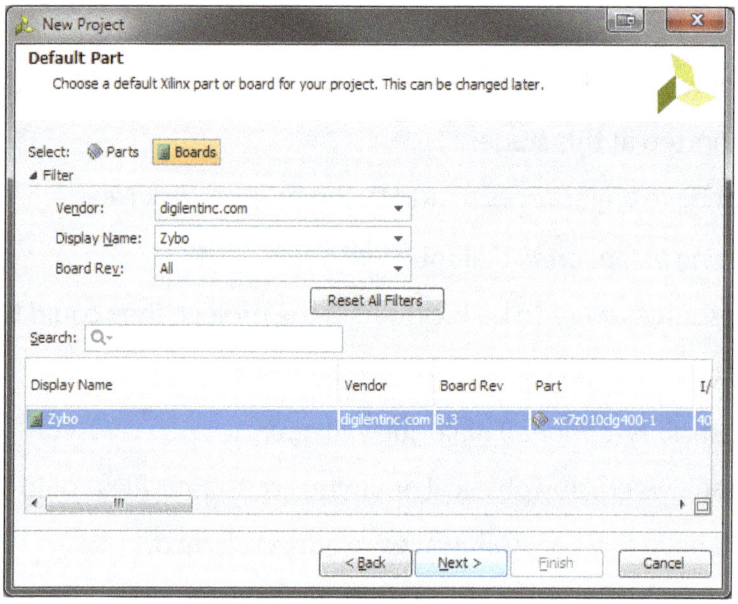

Figure 1.4: Zybo Default Part Dialogue Options

Click ***Next***.

Resume (j) In the *New Project Summary* dialogue, review the specified options, and click ***Finish*** to create the project.

Now that we have created our first project in Vivado IDE, we can now move on to creating our first Zynq embedded system design.

Before doing that, the Vivado IDE tool layout should be introduced. The default Vivado IDE environment layout is shown in Figure 1.5 (other layouts can be chosen by selecting different perspectives). This layout is specifically targeted for the Zedboard. If you are using the Zybo, you will see a slightly different layout.

With reference to the numbered labels in Figure 1.5, the main components of the Vivado IDE environment are:

1. **Menu Bar** - The main access bar gives access to the Vivado IDE commands.
2. **Main Toolbar** - The main toolbar provides easy access to the most commonly used Vivado IDE commands. Tooltips provide information about each command on the toolbar and these can be viewed by hovering the mouse pointer over the buttons, as shown in Figure 1.6.

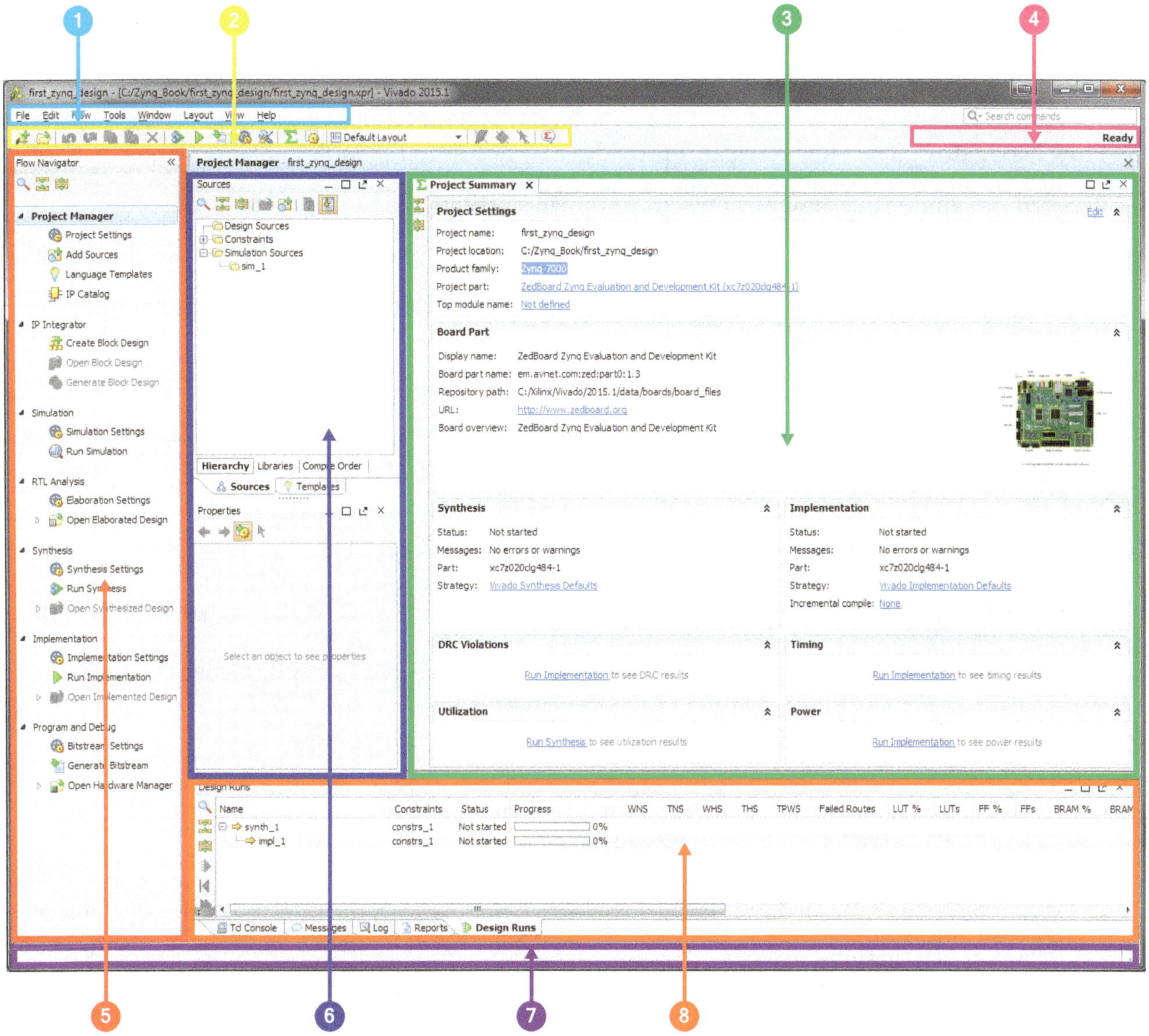

Figure 1.5: Vivado IDE Environment Layout (Zedboard)

Figure 1.6: Toolbar tooltips

3. **Workspace** - The workspace provides a larger area for panels which require a greater screen space and those with a graphical interface, such as:

- Schematic panel
- Device panel
- Package panel

- Text editor panel

4. **Project Status Bar** - The project status bar displays the status of the currently active design.

5. **Flow Navigator** - The Flow Navigator provides easy access to the tools and commands that are necessary to guide your design from start to finish, starting in the *Project Manager* section with design entry and ending with bitstream generation in the *Program and Debug* section. Run commands are available in the *Simulation, Synthesis* and *Implementation* sections to simulate, synthesise and implement the active design.

6. **Data Windows Pane** -The Data Windows pane, by default, displays information that relates to design data and sources, including:

   - **Properties window** - Shows information about selected logic objects or device resources.
   - **Netlist window** - Provides a hierarchical view of the synthesised or elaborated logic design.
   - **Sources window** - Shows IP Sources, Hierarchy, Libraries and Compile Order views.

7. **Status Bar** - The status bar displays a variety of information, including:

   - Detailed information regarding menu bar and toolbar commands will be shown in the lower left side of the status bar when the command is accessed.
   - When hovering over an object in the Schematic window with the mouse pointer, the object details appear in the status bar.
   - During constraint and placement creation in the Device and Package windows, validity and constraint type will be shown on the left side of the status bar. Site coordinates and type will be shown in the right side.
   - The task progress of a running task will be relocated to the right side of the status bar when the **Background** button is selected.

8. **Results Window Area** -The Results Window displays the status and results of commands in a set of windows grouped in the bottom of the Vivado IDE environment. As commands progress, messages are generated and log files and reports are created. The related information is shown here. The default windows are:

   - **Messages** - Displays all messages for the active design.
   - **Tcl Console** - Tcl commands can be entered here and a history of previous commands and outputs are also available.
   - **Reports** - Quick access is provided to the reports generated throughout the design flow.

- **Log** -Displays the log files generated by the simulation, synthesis and implementation processes.
- **Design Runs** -Manages runs for the current project.

Additional windows that can appear in this area as required are: Find Results window, Timing Results window and Package Pins window.

With the layout of the Vivado IDE environment introduced, we can now move on to creating the Zynq system.

Exercise 1B Creating a Zynq System in Vivado

In this exercise we will create a simple Zynq embedded system which implements a General Purpose Input/Output (GPIO) controller in the PL of the Zynq device. The GPIO controller will connect to the LEDs. It will also be connected to the Zynq processor via an AXI bus connection, allowing the LEDs to be controlled by a software application which we will create in Exercise 1C.

A graphical representation of the Zynq embedded design is provided in Figure 1.7.

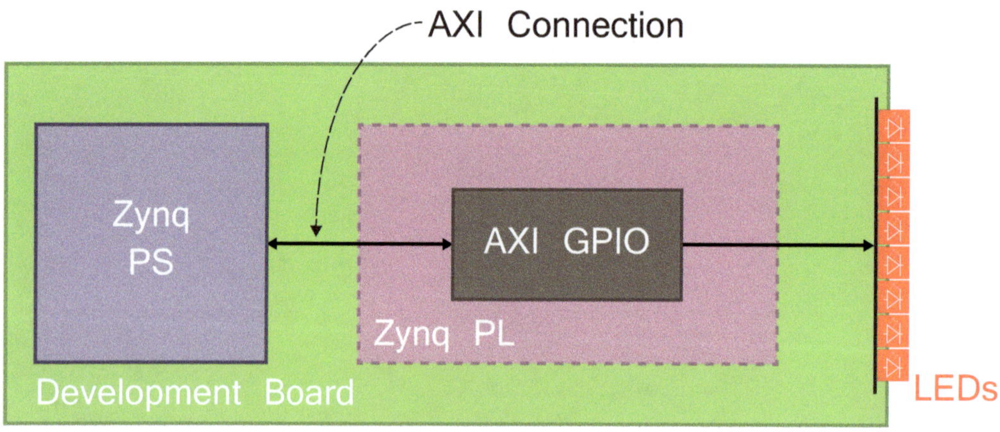

Figure 1.7: Zynq Embedded Design for Exercise 1B

We will begin by creating a new Block Design in Vivado IDE.

(a) In the *Flow Navigator* window, select ***Create Block Design*** from the *IP Integrator* section, as in Figure 1.8:

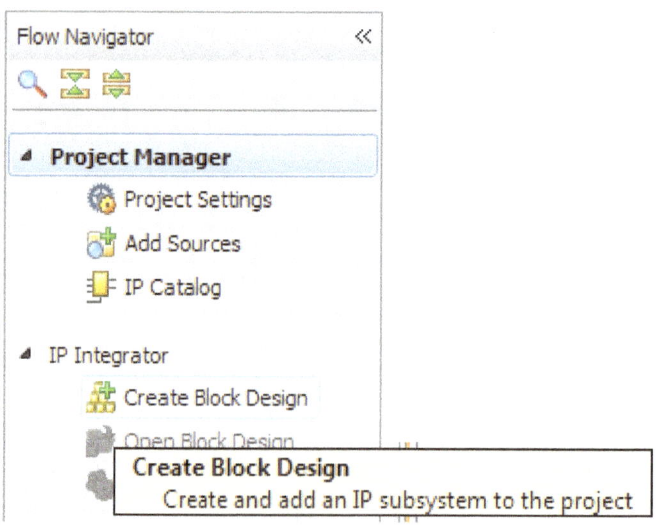

Figure 1.8: Creating a new Block Design in Flow Navigator

The *Create Block Design* dialogue will open.

(b) Enter ***first_zynq_system*** in the Design name box, as in Figure 1.9:

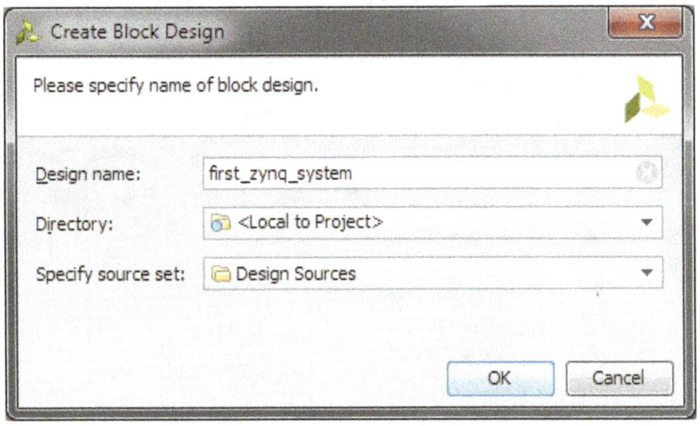

Figure 1.9: Create Block Design dialogue

Click ***OK***. The *Vivado IP Integrator Diagram* canvas will open in the *Workspace*.

The first block that we will add to our design will be a Zynq Processing System.

(c) In the *Vivado IP Integrator Diagram* canvas, right-click anywhere and select ***Add IP***, as in Figure 1.10.

Alternatively, select the ***Add IP*** button in the toolbar at the left of the canvas, shown in Figure 1.11.

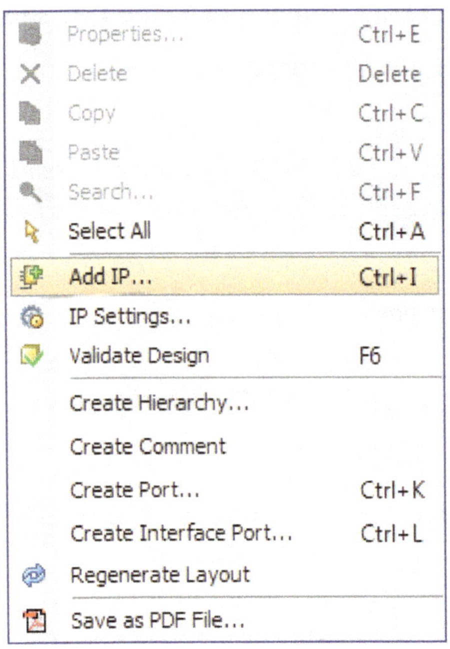

Figure 1.10: Add IP Option

Figure 1.11: Add IP option in IP Integrator canvas information message

The pop-up IP Catalog window will open, as in Figure 1.12.

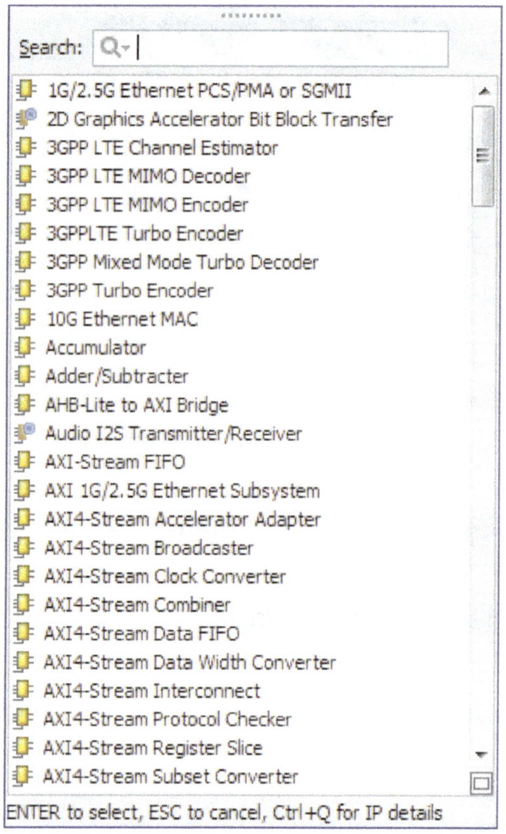

Figure 1.12: Pop-up IP Catalog Window

(d) Enter ***zynq*** in the search field and select the ***ZYNQ7 Processing System***, as shown in Figure 1.13. Be careful not to select the **BFM** version and press the ***Enter*** key on your keyboard.

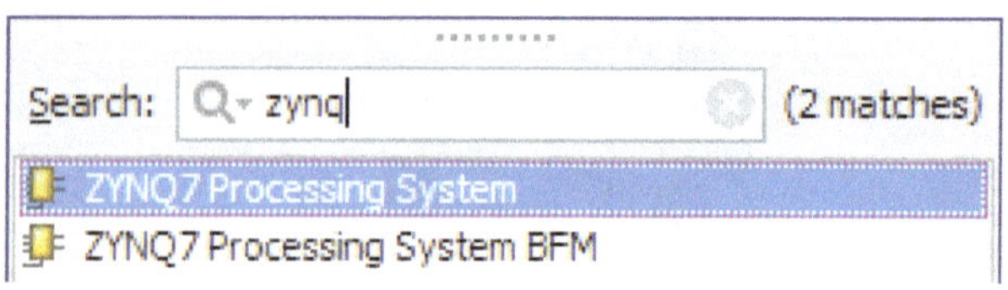

Figure 1.13: Adding ZYNQ7 Processing System from IP CatalogCatalog

You should see a similar message to the following in the *Tcl Console* window to confirm that the processing system has indeed been added to the design correctly:

```
create_bd_cell -type ip -vlnv xilinx.com:ip:processing_system7:5.5
processing_system7_0
```

Messages like this will be displayed in the *Tcl Console* window for all actions carried out on IP Integrator blocks.

The next step is to connect the **DDR** and **FIXED_IO** interface ports on the Zynq PS to the top-level interface ports on the design.

(e) Click the ***Run Block Automation*** option from the *Designer Assistance* message at the top of the Diagram window, as shown in Figure 1.14.

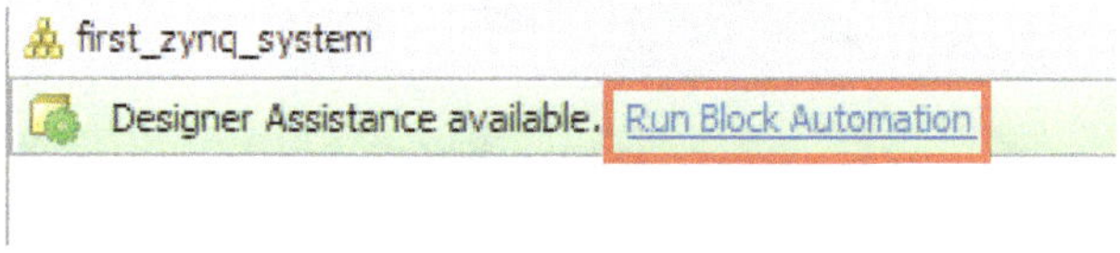

Figure 1.14: Run Block Automation - Processing System

Zed

In the *Run Block Automation* dialogue, ensure that the option to ***Apply Board Preset*** is selected and click ***OK***. The external connections for both the **DDR** and **FIXED_IO** interfaces will now be generated.

Your block diagram should now resemble Figure 1.15.

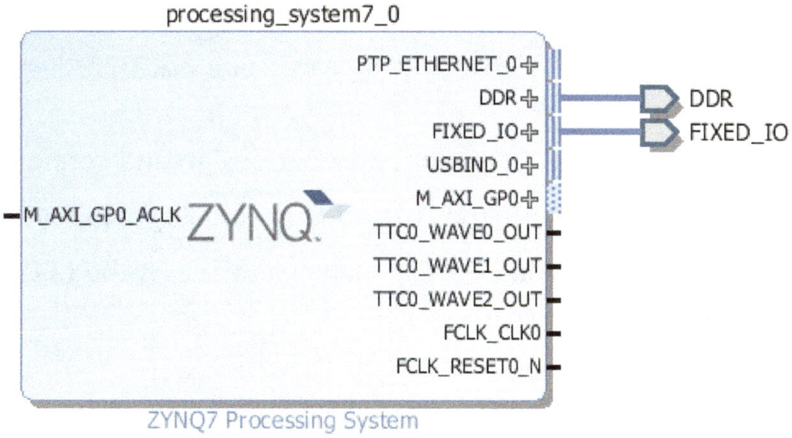

Figure 1.15: Zedboard ZYNQ7 Processing System External Connections

As the ***ZedBoard*** platform is the target development board, and this was specified on creation of the project, Vivado will configure the Zynq processor block accordingly.

Zybo

In the *Run Block Automation* dialogue, ensure that the option to ***Apply Board Preset*** is selected and click ***OK***. The external connections for both the **DDR** and **FIXED_IO** interfaces will now be generated.

Your block diagram should now resemble Figure 1.16.

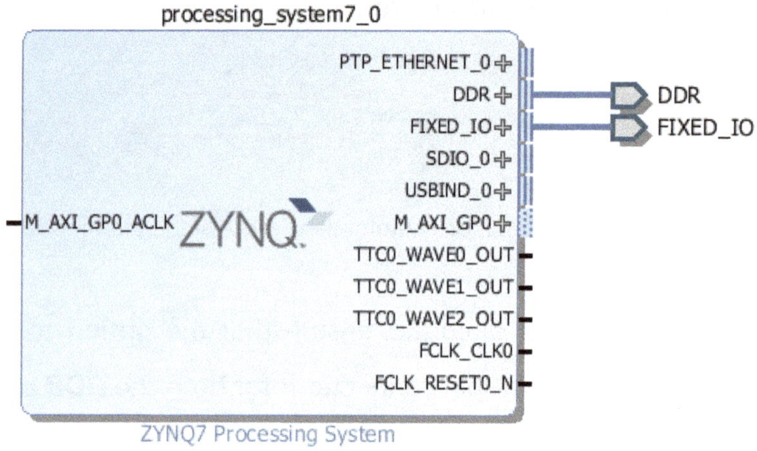

Figure 1.16: Zybo ZYNQ7 Processing System External Connections

As the ***Zybo*** platform is the target development board, and this was specified on creation of the project, Vivado will configure the Zynq processor block accordingly.

Resume Now that the main Zynq PS has been added to our design and configured, we can now add further blocks which will be placed in the PL to add functionality to the system. In this case we will only be adding a single block, **AXI GPIO**, to allow us to access the **LEDs** on the development board.

(f) Right-click in an empty area of the *Diagram* window and select ***Add IP***. Enter ***GPIO*** in the search field and add an instance of the ***AXI GPIO*** IP.

We will now use the *IP Integrator Designer Assistance* tool to automate the connection of the **AXI GPIO** block to the **ZYNQ7 Processing System**.

(g) Click ***Run Connection Automation*** from the *Designer Assistance* message at the top of the *Diagram* window and select ***/axi_gpio_0/S_AXI***, as shown Figure 1.17.

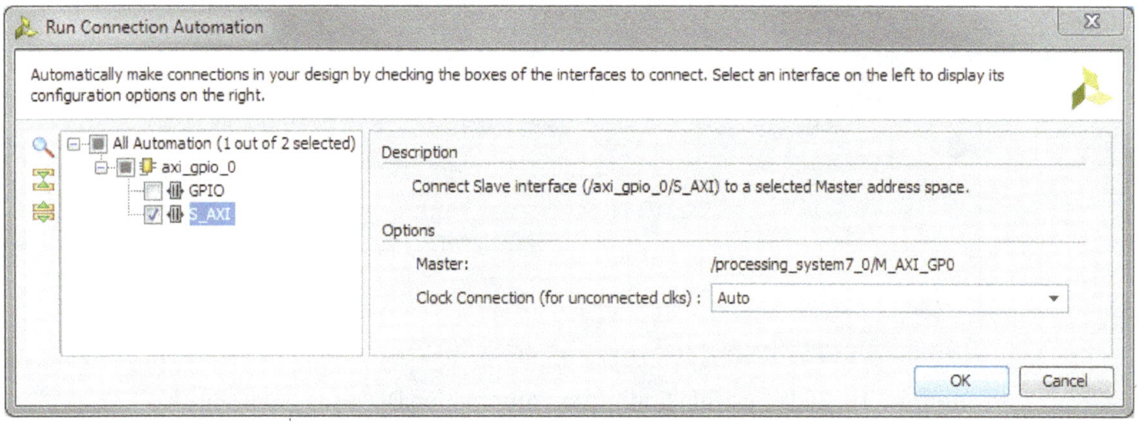

Figure 1.17: Run Block Automation - GPIO Block

This will automate the process of connecting the GPIO to an AXI port, and will automatically instantiate two further IP blocks:

- **Processor System Reset Module** - This provides customised resets for an entire processing system, including the peripherals, interconnect and the processor itself.
- **AXI Interconnect** - Provides an AXI interconnect for the system, allowing further IP and peripherals in the PL to communicate with the main processing system.

Click ***OK***.

All connections between the blocks should be made automatically.

(h) One final connection is required to connect the **AXI GPIO** block to the **LEDs** on the development board. This can also be completed using *Designer Assistance.*

Zed

Click ***Run Connection Automation*** from the *Designer Automation* message at the top of the *Diagram* window. The Run Connection Automation dialogue will open, as shown in Figure 1.18.

Select ***/axi_gpio_0/GPIO***.

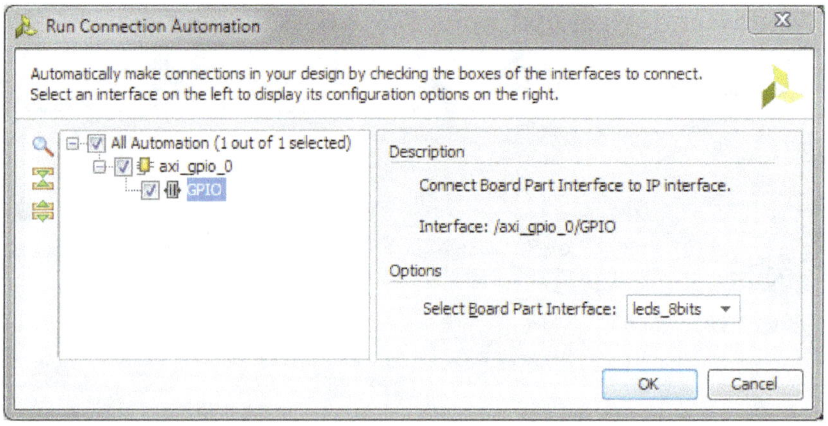

Figure 1.18: Zedboard Run Connection Automation Dialogue - GPIO Block

Select ***leds_8bits*** from the *Select Board Part Interface* drop-down menu, and click ***OK***.

The gpio interface of the AXI GPIO block will now be connected to the LEDs on the development board, and your complete design should resemble Figure 1.19.

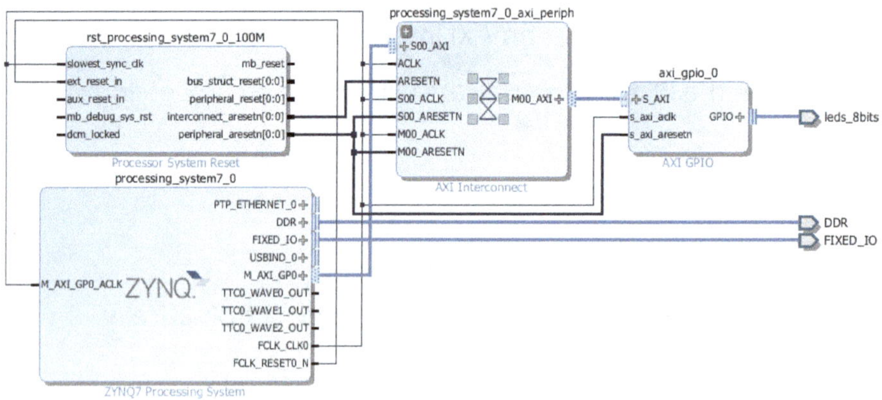

Figure 1.19: Zedboard, Zynq Processor System

The positions of the individual IP blocks in your design may vary slightly from Figure 1.19, but the blocks and their connections should be the same.

Zybo Click ***Run Connection Automation*** from the *Designer Automation* message at the top of the *Diagram* window. The Run Connection Automation dialogue will open, as in Figure 1.20.

Select ***/axi_gpio_0/GPIO***.

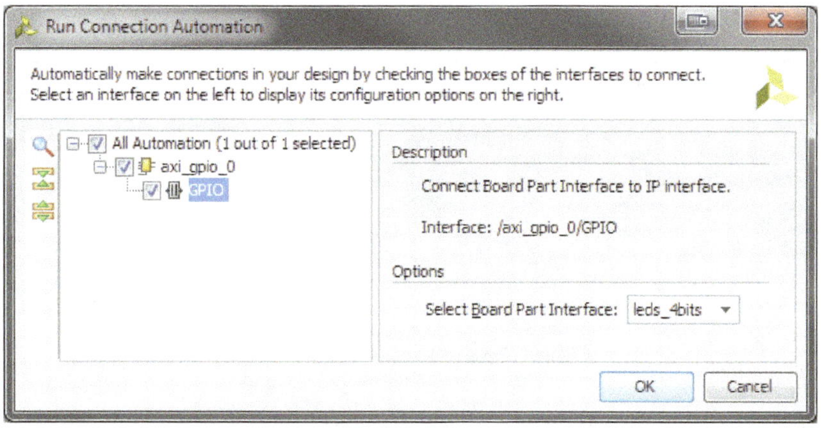

Figure 1.20: Zybo Run Connection Automation Dialogue - GPIO Block

Select ***leds_4bits*** from the *Select Board Part Interface* drop-down menu, and click ***OK***.

The gpio interface of the AXI GPIO block will now be connected to the LEDs on the development board, and your complete design should resemble Figure 1.21.

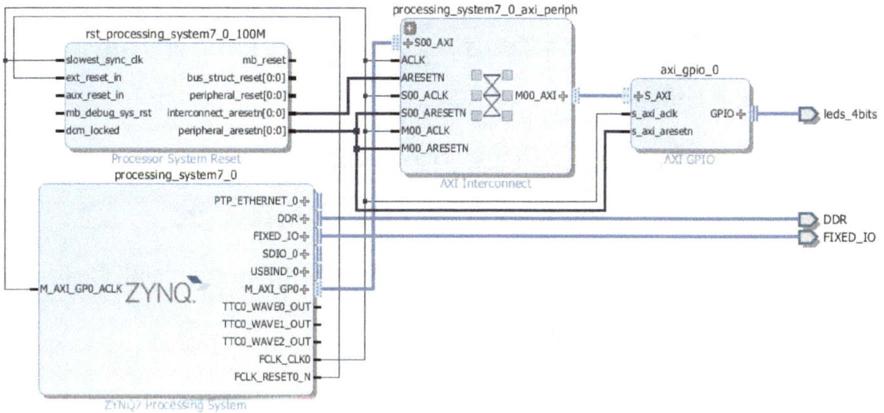

Figure 1.21: Zybo, Zynq Processor System

The positions of the individual IP blocks in your design may vary slightly from Figure 1.21, but the blocks and their connections should be the same.

Resume IP Integrator will automatically assign a memory map for all IP that is present in the design. We will not be changing the memory map in this tutorial, but for future reference we will take a look at the Address Editor.

(i) Select the Address Editor tab from the top of the Workspace window, as shown in Figure 1.22, and expand the Data group.

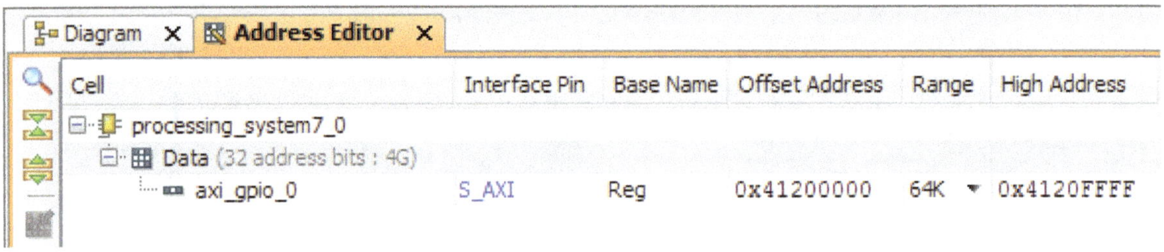

Figure 1.22: Address Editor Tab

You can see that IP Integrator has already assigned a memory map (the mapping of specific sections of memory to the memory-mapped registers of the IP blocks in the PL) to the AXI GPIO interface, and that it has a range of **64K**.

Now that our system is complete, we must first validate the design before generating the HDL design files.

(j) Save your design by selecting ***File > Save Block Design*** from the *Menu Bar*.

(k) Validate the design by selecting ***Tools > Validate Design*** from the *Menu Bar*. This will run a Design-Rule-Check (DRC).
Alternatively, select the Validate Design button, , from the Main Toolbar, or right-cick anywhere in the *Diagram* canvas and select ***Validate Design***.

(l) A *Validate Design* dialogue should appear to confirm that validation of the design was successful. Click ***OK*** to dismiss the message.

With the design successfully validated, we can now move on to generating the HDL design files for the system.

(m) Switch to the **Sources Tab** by selecting ***Window > Sources*** from the *Menu Bar*.

(n) In the **Sources** window, right-click on the top-level system design, which in this case is **first_zynq_system**, and select ***Create HDL Wrapper***, as shown in Figure 1.23.

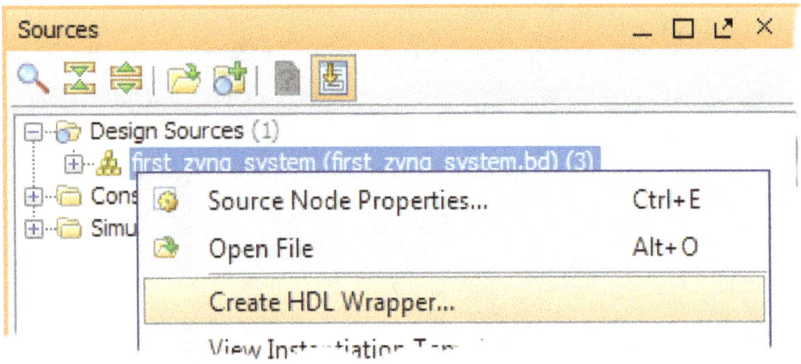

Figure 1.23: Create HDL Wrapper

The *Create HDL Wrapper* dialogue window will open. Select *Let Vivado manage wrapper and auto-update*, and click ***OK***.

This will generate the top level HDL wrapper for our system.

All of the source files for the IP blocks that were used in the IP Integrator block diagram, as well as any relevant constraints files, will be generated during the synthesis process. As we specified VHDL as the target language when creating the project in Exercise 1A, all generated source files will be VHDL.

With all HDL design files generated, the next step in Vivado is to implement our design and generate a bitstream file.

(o) In *Flow Navigator*, click ***Generate Bitstream*** from the *Program and Debug* section. If a dialogue window appears prompting you to save your design, click ***Save***.

(p) A dialogue window will open requesting that you launch synthesis and implementation before starting the *Generate Bitstream* process. Click ***Yes*** to accept.

The combination of running the synthesis, implementation and bitstream generation processes back-to-back may take a few minutes, depending on the power of your computer system.

(q) Once the bitstream generation is finished, a dialogue window will open to inform you that the process has been completed successfully, as in Figure 1.24.

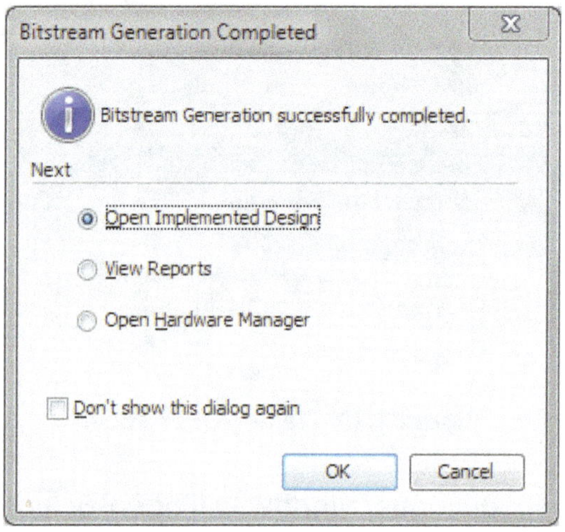

Figure 1.24: Bitstream Generation Completed Dialogue Window

Select ***Open Implemented Design***, and click ***OK***.

At this point you will be presented with the *Device* view, where you can see the PL resources that are utilised by the design. With the default colour scheme, these are shown in light blue.

With the bitstream generated, the building of the hardware image is complete. It must now be exported to a software environment where we will build a software application to control and interact with the custom hardware.

The final step in Vivado is to export the design to the SDK, where we will create the software application that will allow the Zynq PS to control the LEDs on the development board.

(r) Select ***File > Export > Export Hardware...*** from the *Menu Bar*.

(s) The *Export Hardware* dialogue window will open. Ensure that the option to ***Include bitstream*** is selected, as in Figure 1.25, and Click ***OK***.

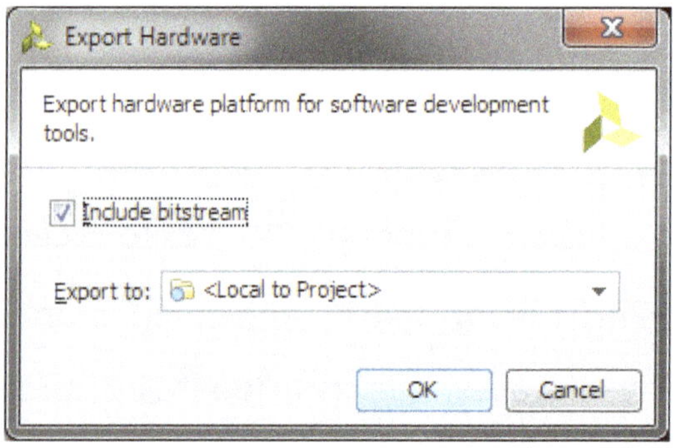

Figure 1.25: Export Hardware for SDK

NOTE: For the option to *Include bitstream* to be enabled, an implemented design must be active. This is the reason that we opened the implemented design in Step (q).

(t) Launch the SDK in Vivado by selecting ***File > Launch SDK*** from the *Menu Bar* and Click ***OK***.

This concludes the steps that are required in Vivado IDE. All hardware components of the system have been configured and generated. In the next exercise we will move on to creating a simple software component which will control the system.

Exercise 1C Creating a Software Application in the SDK

In this exercise we will create a simple software application which will control the LEDs on the Zynq development board. The software application will run on the Zynq processing system and communicate with the AXI GPIO block which is implemented in the PL. We will take a look at the software drivers that are created by IP Integrator for each of the IP modules, before building and executing the software on the development board.

The SDK should have opened after the conclusion of Exercise 1B. If it did not open, you can open the SDK by navigating to **Start > All Programs > Xilinx Design Tools > SDK 2015.1 > Xilinx SDK 2015.1**

When launching the SDK from the start menu, you will need to specify the workspace that was created when the Vivado IP Integrator design was exported in Exercise 1B. It should be:

C:\Zynq_Book\first_zynq_design\first_zynq_design.sdk

Enter this in the *Workspace* field of the *Workspace Launcher* dialogue window, as shown in Figure 1.26.

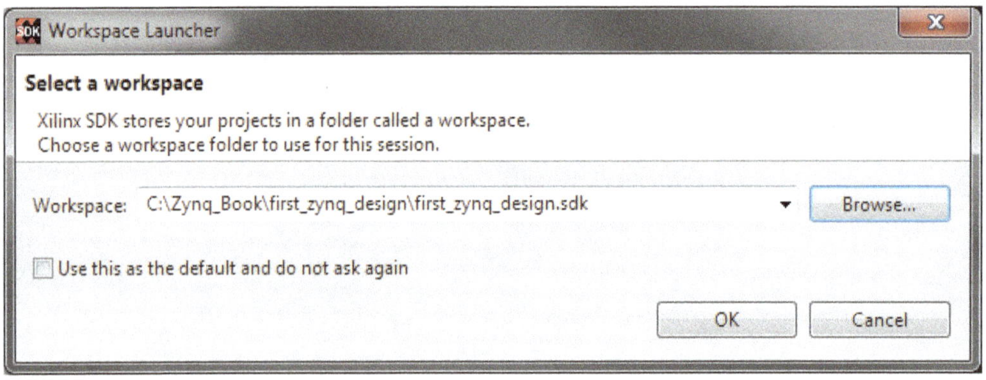

Figure 1.26: SDK Workspace Launcher Dialogue Window

With the SDK open, we can begin the creation of our software application. You will already be able to see the Hardware Platform Project, which will be automatically created and opened. It is now necessary to add an Application Project and a Board Support Package.

(a) Select **File > New > Application Project** from the *Menu bar*.

(b) The *New Project* dialogue window will open. Enter ***LED_test*** in the *Project name* field, as shown in Figure 1.27, keeping all other options with the default settings. Click ***Next*** (Be careful not to select *Finish*).

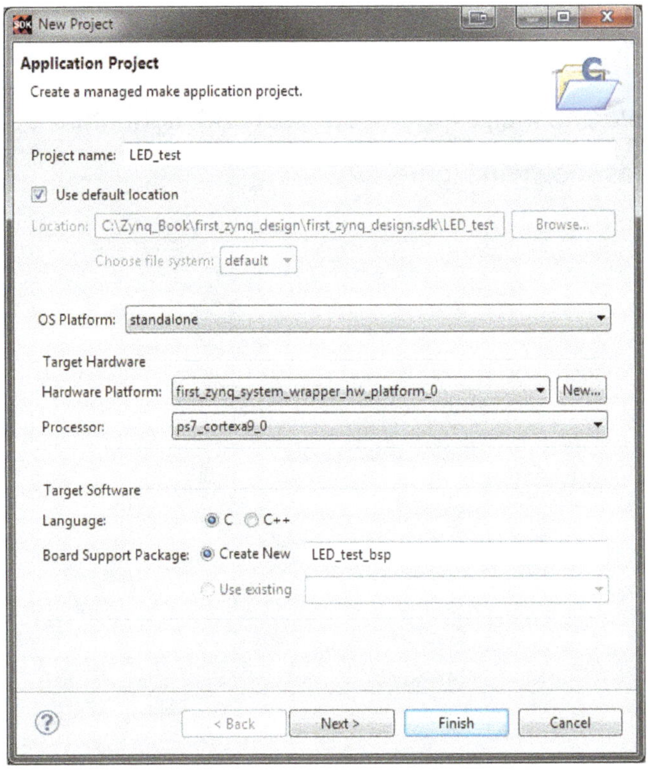

Figure 1.27: New Application Project Dialogue

(c) At the *New Project Templates* screen, select ***Empty Application***, as in Figure 1.28, and click ***Finish*** to create the project.

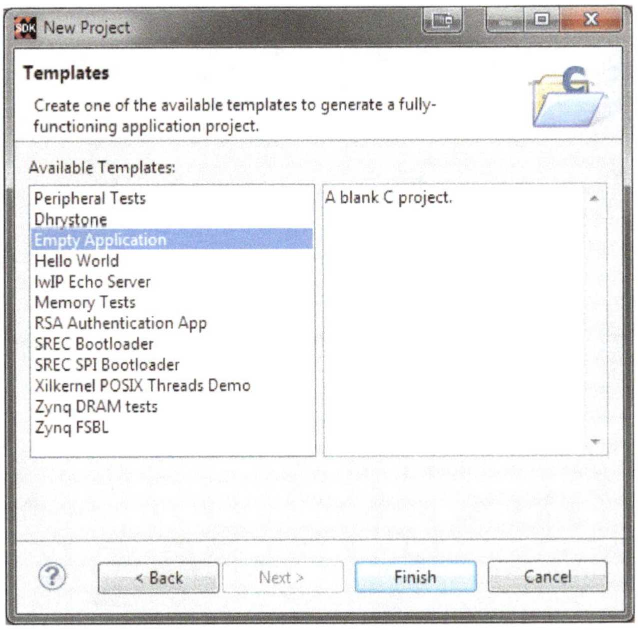

Figure 1.28: New Project Template Dialogue

NOTE: the new project should open automatically. If it doesn't, you may need to close the Welcome tab in order to view the project.

With the new Application Project created, we can now import some pre-prepared source code for the application.

(d) In the *Project Explorer* panel, expand **LED_test** and highlight the *src* directory. Right-click and select ***Import...***, as shown in Figure 1.29.

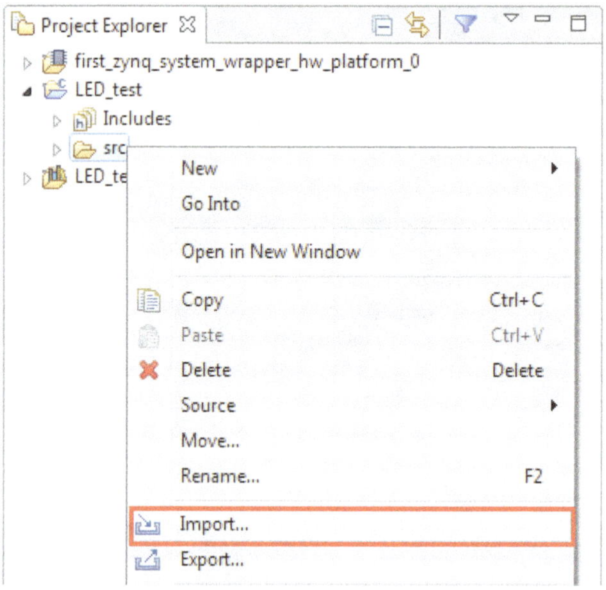

Figure 1.29: Import Source Files to Project

(e) The *Import* window will open. Expand the **General** option and highlight ***File System***, as in Figure 1.30, and click ***Next***.

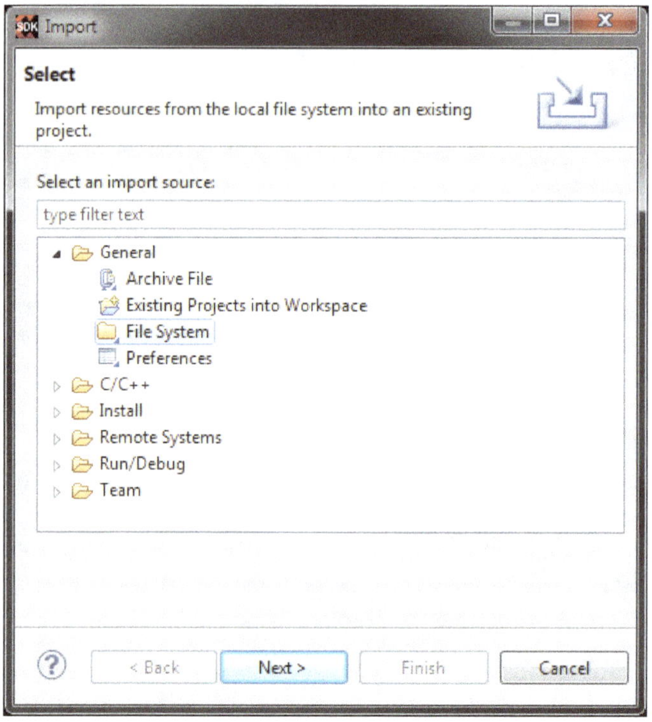

Figure 1.30: Import File System

(f) In the *Import File System* window, click the ***Browse...*** button.

The source file directory will depend on the Zynq development board that is in use. If you are using the **Zedboard**, navigate to: **C:\Zynq_Book\sources\zedboard\first_zynq_design.** If you are using the **Zybo**, navigate to **C:\Zynq_Book\sources\zybo\first_zynq_design**

Click ***OK.***

(g) Select the file **LED_test_tut_1C.c**, as shown in Figure 1.31, and click ***Finish***.

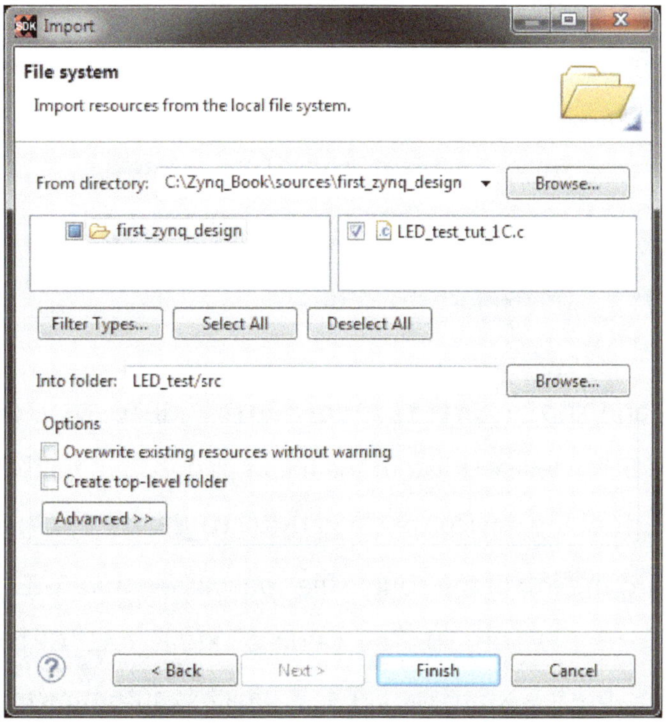

Figure 1.31: Import C Source File

The C source file will be imported and the project should automatically build. You should see a similar message to Figure 1.32 in the *Console* window.

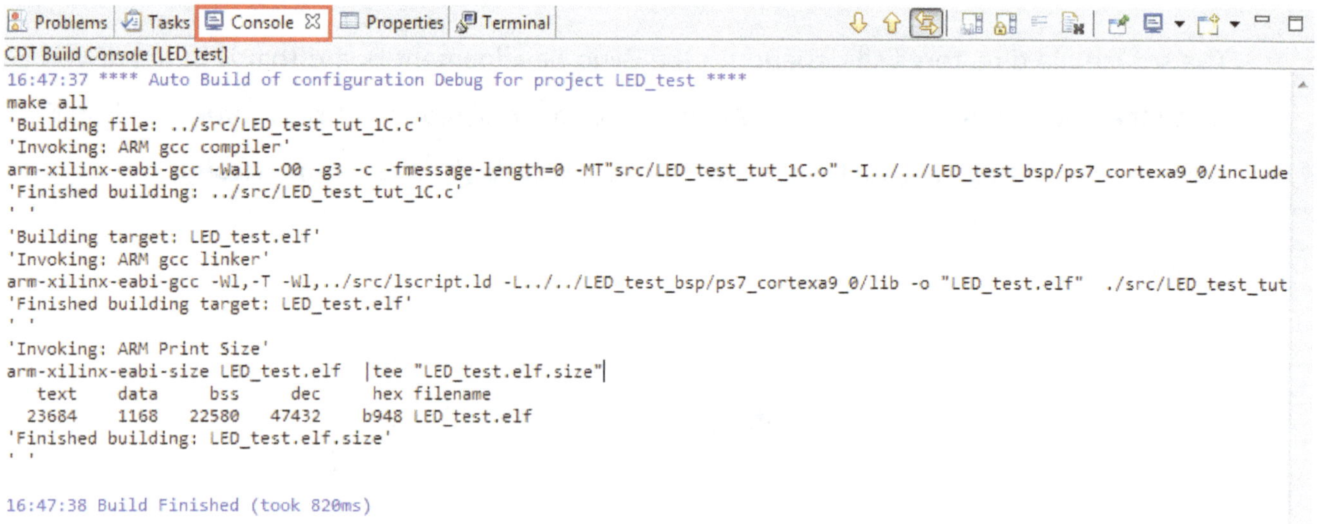

Figure 1.32: Build Finished Console Message

(h) Open the imported source file by expanding the *src* folder and double-clicking on **LED_test_tut_1C.c**, and explore the code.

Note the command `XGpio_Initialize(&Gpio, GPIO_DEVICE_ID);` This is a function provided by the GPIO device driver in the file `xgpio.h`. It initialises the XGpio instance, *Gpio*, with the unique ID of the device specified by `GPIO_DEVICE_ID`.

If you look toward the top of the source file you will see that `GPIO_DEVICE_ID` is defined as `XPAR_AXI_GPIO_0_DEVICE_ID`. The value of `XPAR_AXI_GPIO_0_DEVICE_ID` can be found by opening the file, `xparameters.h`, which is automatically generated by Vivado IDE when exporting a hardware design to the SDK. It contains definitions of all the hardware parameters of the system.

The function, `XGpio_SetDataDirection(&Gpio, LED_CHANNEL, 0xFF);` is also provided by the GPIO device driver, and sets the direction of the specified GPIO port. As we are specifying the LEDs in this case, it is specifying an output. Bits set to '0' are output, and bits set to '1' are input. As there are 8 LEDs on the **Zedboard**, by setting the LED channel direction to a value of `0x00`, or `00000000` in binary, we are setting all 8 LEDs as outputs. Similarly as there are 4 LEDs on the **Zybo** board, by setting the LED channel direction a value of `0x0` or `0000` in binary, we are setting all 4 LEDs as outputs.

Further information on the peripheral drivers can be found by selecting the **system.mss** tab. A list of all the peripherals in the system is provided, along with links to available documentation and examples, as shown in Figure 1.33.

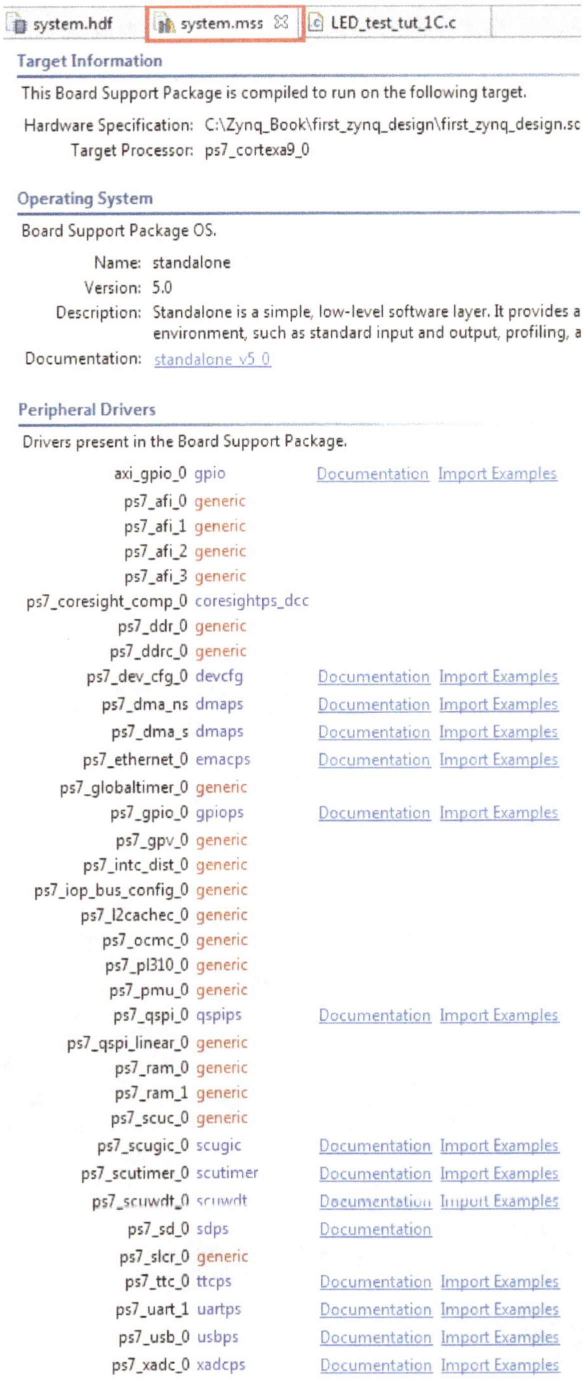

Figure 1.33: Peripheral Documentation and Drivers in system.mss tab

The next step is to program the Zynq PL with the bitstream file that we generated in Exercise 1B.

(i) Ensure that the Zynq development board is powered on and that the JTAG port is connected to the PC via the provided USB-A to USB-B cable. Additionally the board jumpers must also be correctly set so that to enable JTAG mode, which allows the hardware to be programmed and access for system debugging tools.

Zed

The **Zedboard** requires five jumpers to be set as shown in Figure 1.34. This configuration will enable JTAG mode.

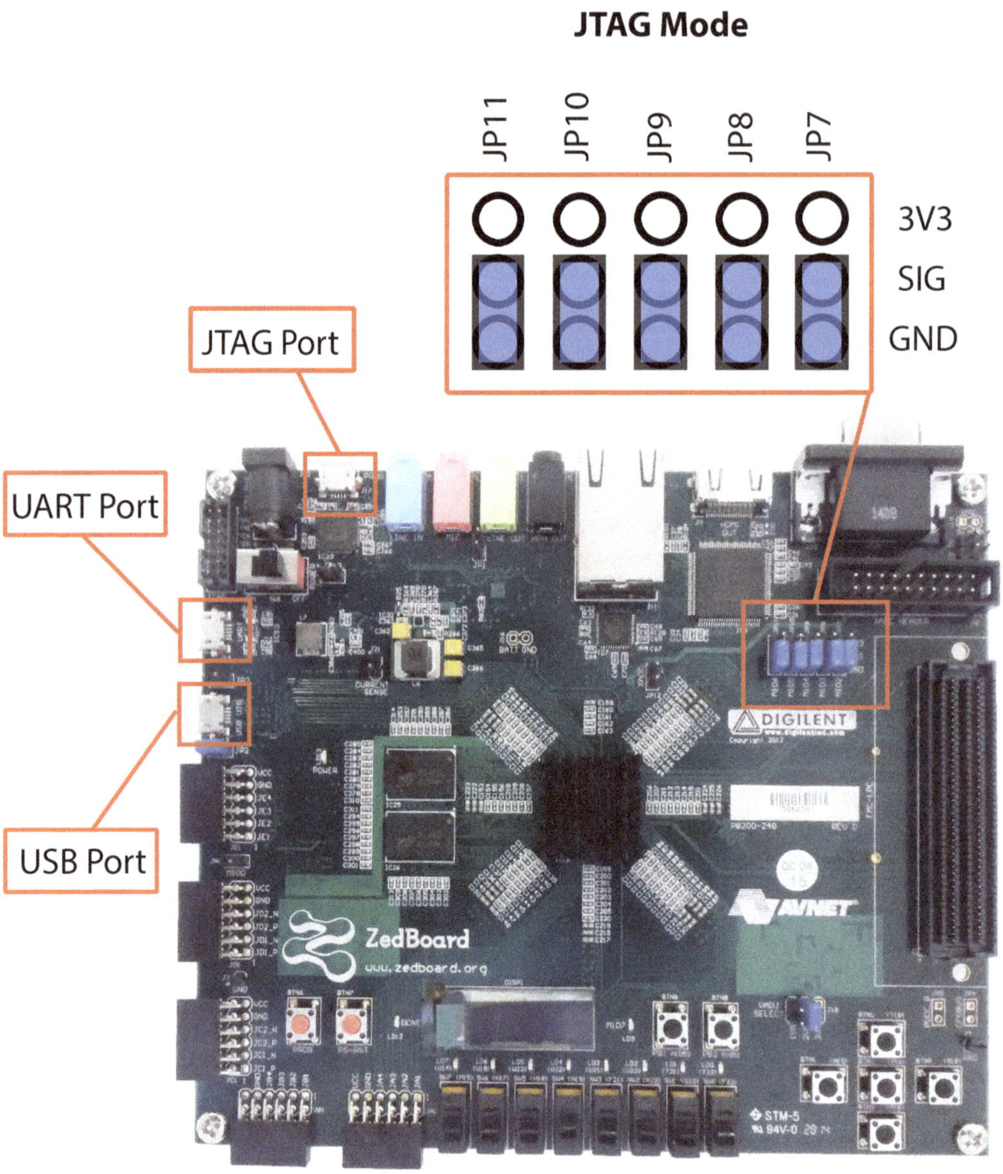

Figure 1.34: Zedboard JTAG Jumper Configuration

Zybo

One jumper is set to enable JTAG mode on the **Zybo** development board. Additionally the board's power supply is set using a jumper with the possibility of receiving power from USB, wall or battery. Both JTAG and power jumper configurations are set in Figure 1.35. The board has been set to receive power from USB.

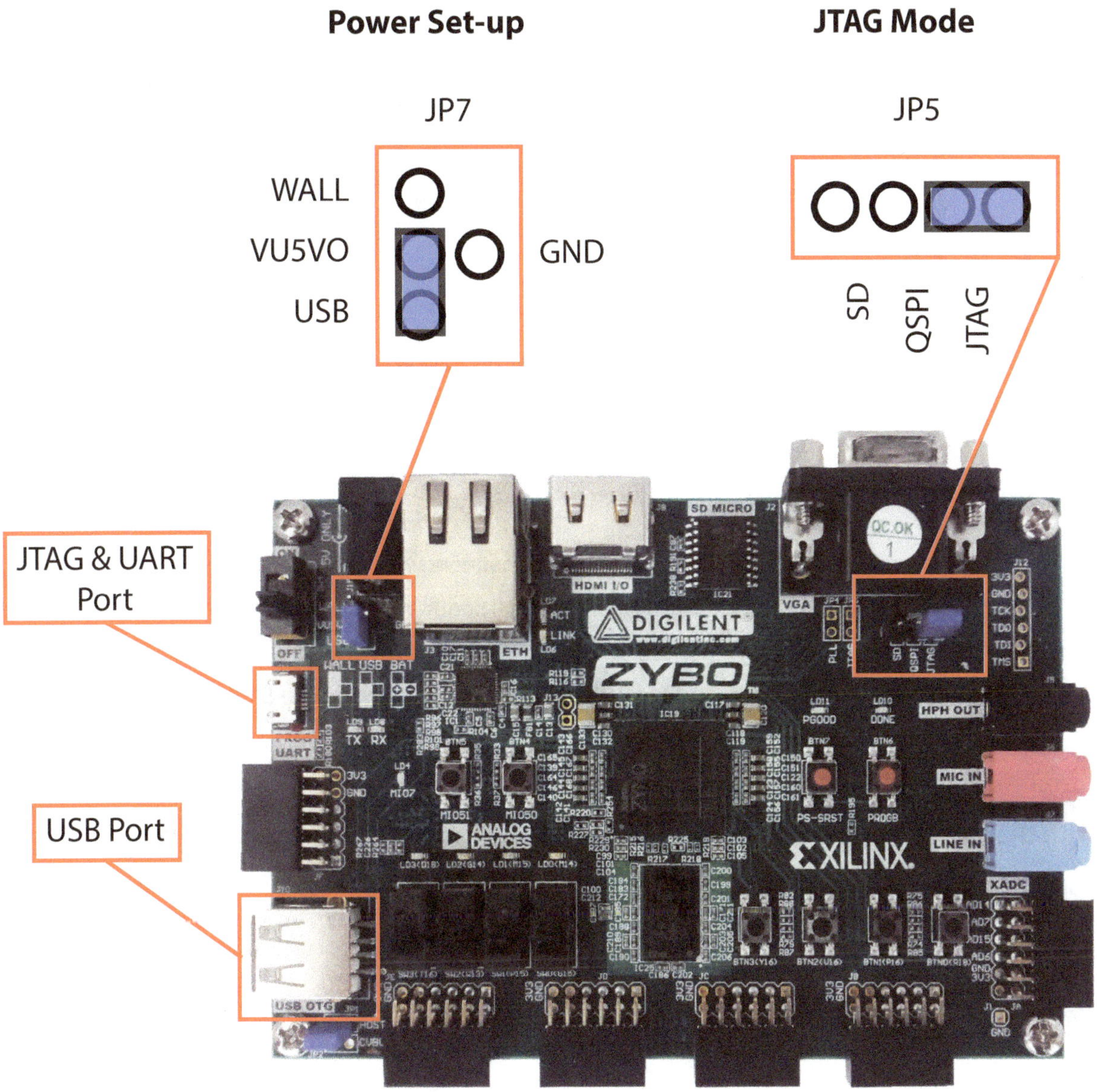

Figure 1.35: Zybo JTAG and Power Jumper Configurations

Resume (j) Download the bitstream to the Zynq PL by selecting **Xilinx Tools > Program FPGA** from the *Menu bar*. The *Program FPGA* window will appear. The Bitstream field should already be populated with the correct bitstream file, as in Figure 1.36.

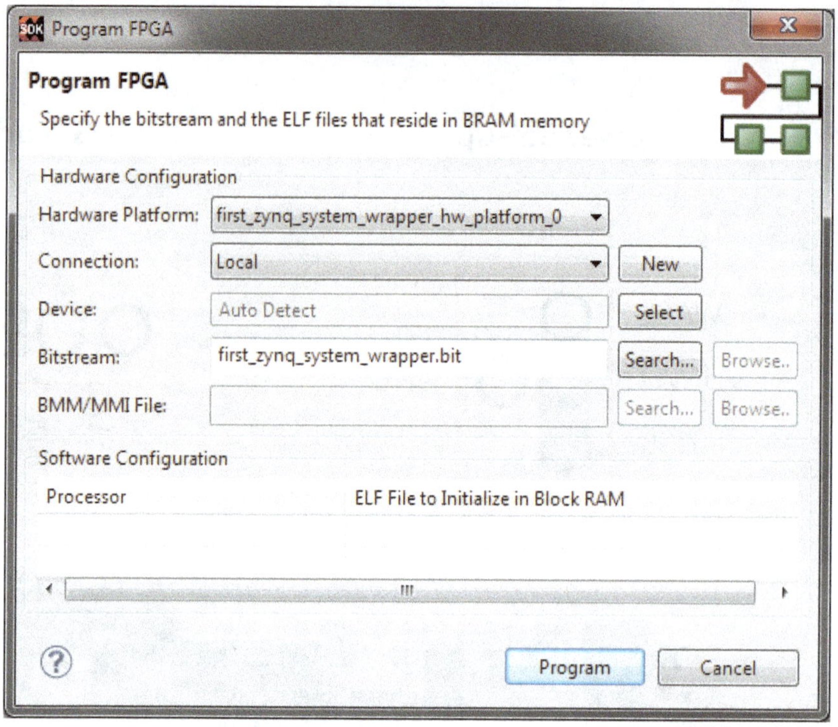

Figure 1.36: Program FPGA dialogue Window

NOTE: Once the device has successfully been programmed, the *DONE LED* on the ***ZedBoard*** will turn blue. Similarly the *DONE LED* on the ***Zybo*** will turn green.

With the Zynq PL successfully configured with the bitstream file, we can now launch our software application on the Zynq PS.

(k) Select the project ***LED_test*** in *Project Explorer*. Right-click and select **Run As > Launch on Hardware (GDB)** as in Figure 1.37.

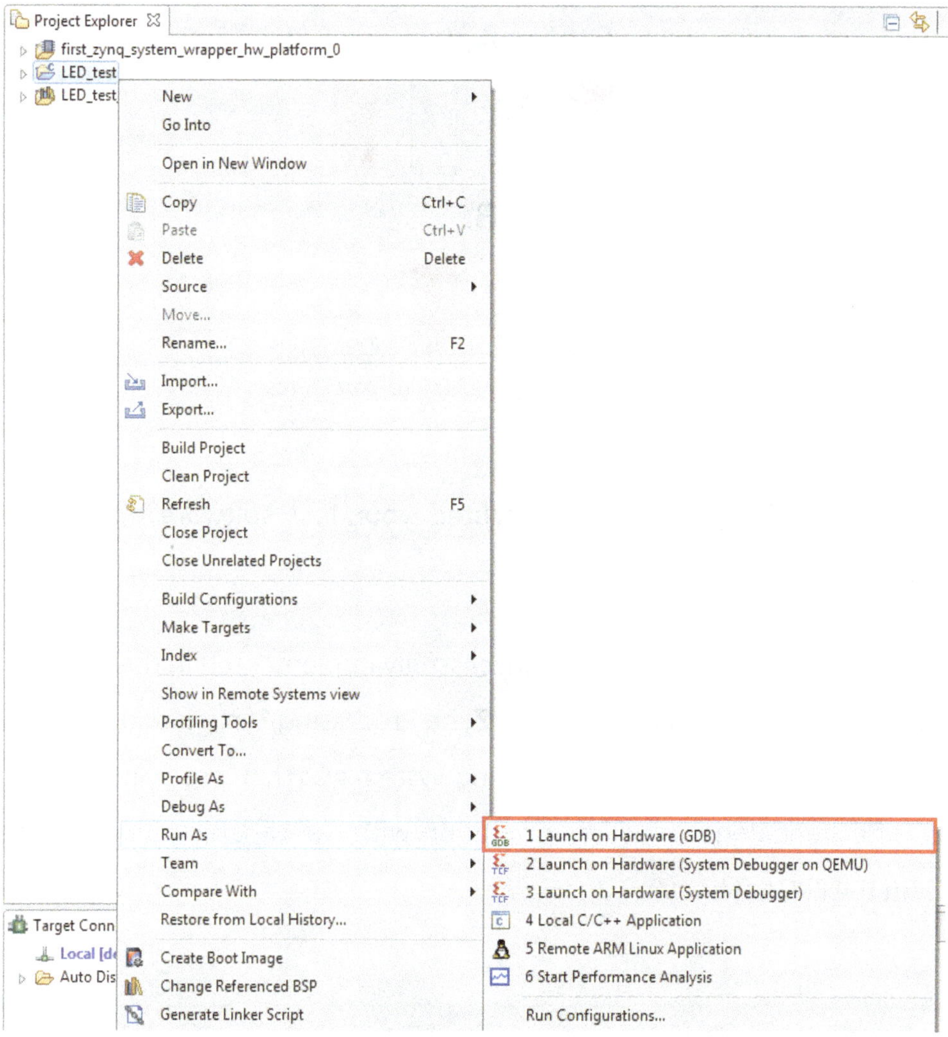

Figure 1.37: Launch Application onto the Zynq Development Board

After a few seconds the LEDs on the **ZedBoard** should begin to flash between the states highlighted in Figure 1.38.

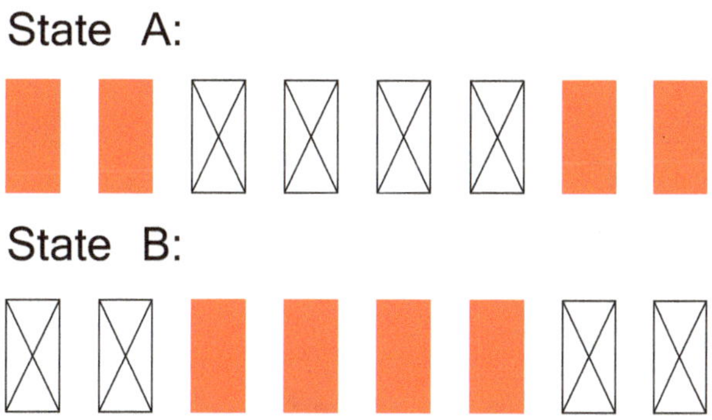

Figure 1.38: Zedboard LED Flashing States

Zybo The LEDs on the **Zybo** should begin to flash between the states highlighted in Figure 1.39

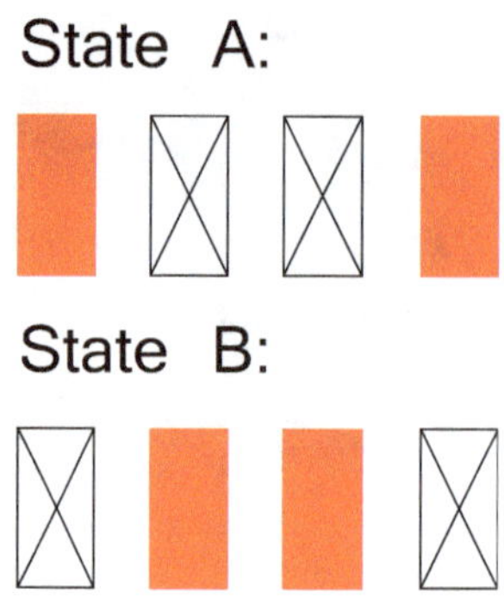

Figure 1.39: Zybo LED Flashing States

Resume You have successfully created and executed your first software application on the Zynq processing system.

In summary a GPIO controller has been successfully implemented in the FPGA fabric of the Zynq device, forming a connection between the Zynq Processing System and the development board LEDs via an AXI interface. The Zynq Processing System was then programmed to control the LEDs by means of a standalone software application with the capability to interface with the GPIO controller in the FPGA fabric.

The Zynq Book Tutorials 2

Next Steps in Zynq SoC Design

v1.4, June 2015

Revision History

Date	Version	Changes
13/09/2013	1.0	First release for Vivado Design Suite version 2013.2
27/01/2014	1.1	Updated for changes in Vivado Design Suite version 2013.4
30/04/2014	1.2	Updates for changes in Vivado Design Suite version 2014.1
08/04/2015	1.3	Updated for changes in Vivado Design Suite version 2014.4
29/04/2015	1.3.1	Updated to include Zybo development board for Vivado Design Suite version 2014.4
19/06/2015	1.4	Updated for changes in Vivado Design Suite version 2015.1

Introduction

This tutorial will guide you through the process of creating a Zynq design utilising interrupts. Using the Vivado™ Integrated Development Environment (IDE) and the IP Integrator environment, a simple Zynq™ processor design, to be implemented on a Zynq development board, will be generated. The Software Development Kit (SDK) will then be used to create a simple software application which will run on the Zynq's ARM Processing System (PS) to control the hardware that is implemented in the Programmable Logic (PL). This tutorial leads on from the previous one, expanding on the skills acquired in it.

The tutorial is split into four exercises, and is organised as follows:

Exercise 2A — This exercise provides a further guide to the process of launching Vivado IDE and creating a project using *New Project Wizard.*

Exercise 2B — In this exercise, we will use the project that was created in Exercise 2A to build a Zynq embedded system utilising interrupts with IP Integrator and incorporating existing IP from the Vivado IP Catalog. This will expand on previous knowledge gained in creating and connecting a block based system in IP Integrator. The completed design will have an associated bitstream generated and will be exported to the Xilinx SDK to participate in a test application.

Exercise 2C — In the Xilinx SDK, a test software application for the generated hardware system will be created and explained. Running this application on a Zynq development board will demonstrate the function of interrupts and how the application is coded to utilise them.

Exercise 2D — Finally, we will return to the system from Exercise 2B and include an additional source of interrupt, making the necessary connections, generating a bitstream and exporting to the Xilinx SDK. We will then modify our previous software application to inspect the operation of the altered system.

Exercise 2A Expanding the Basic IP Integrator Design

In this exercise we will expand upon the previous project in Vivado IDE by adding additional GPIO and configuring the system to utilise interrupts. For the sake of clarity and understanding, we will run through the building of a basic system once more. Start by launching the Vivado IDE.

(a) Launch Vivado by double-clicking on the Vivado desktop icon: , or by navigating to **Start > All Programs > Xilinx Design Tools > Vivado 2015.1 > Vivado 2015.1**

(b) When Vivado loads, you will be presented with the *Getting Started* screen as in Figure 2.1.

Figure 2.1: Vivado IDE Getting Started screen

(c) Select the option to ***Create New Project*** as in Figure 2.2.

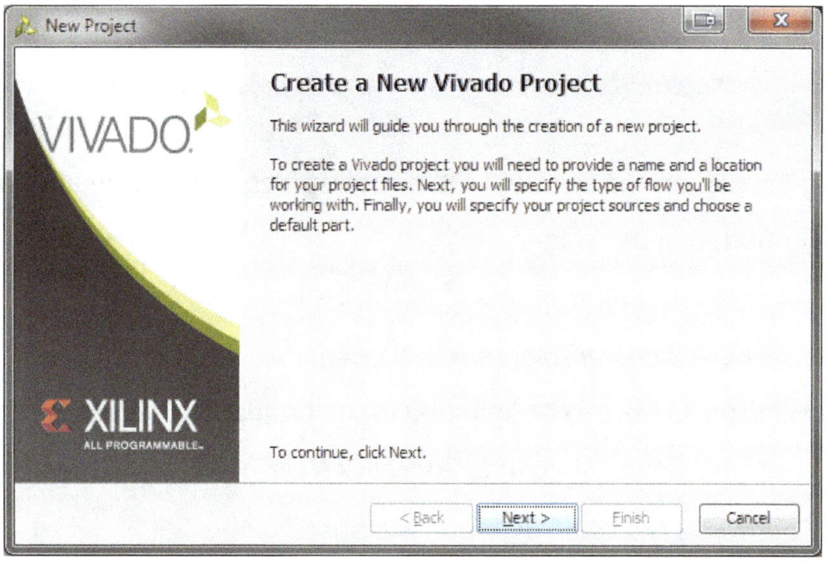

Figure 2.2: New Project dialogue

Click ***Next***.

(d) At the Project Name dialogue, enter ***zynq_interrupts*** as the ***Project name*** and ***C:/Zynq_Book*** as ***Project location***.

Make sure that you select the option to ***Create project subdirectory***. All options should be the same as shown below:

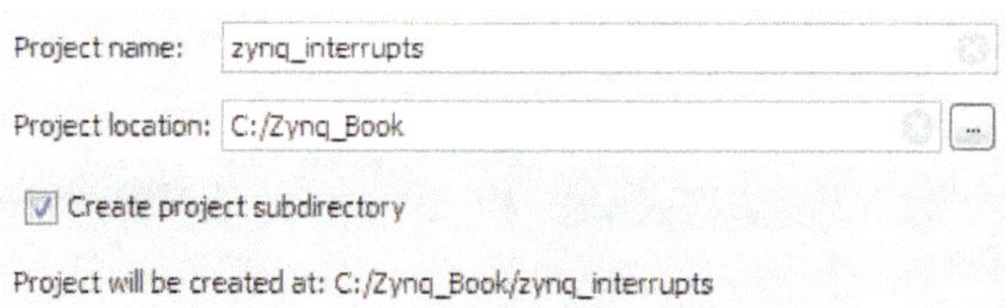

Click ***Next***.

A directory named ***Zynq_Book*** will be created on your ***C drive*** if it did not already exist.

(e) At the *Project Type* dialogue, select ***RTL Project*** and ensure that the option ***Do not specify sources at this time*** is not selected:

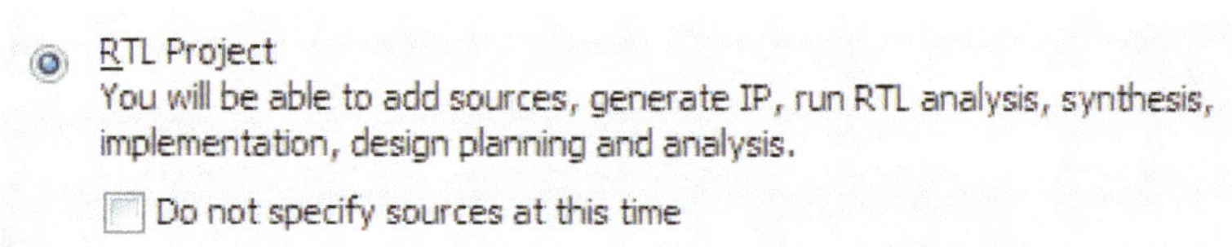

Click ***Next***.

(f) Select ***VHDL*** as the ***Target language*** and ***Mixed*** as the ***Simulator Language*** in the *Add Sources* dialogue:

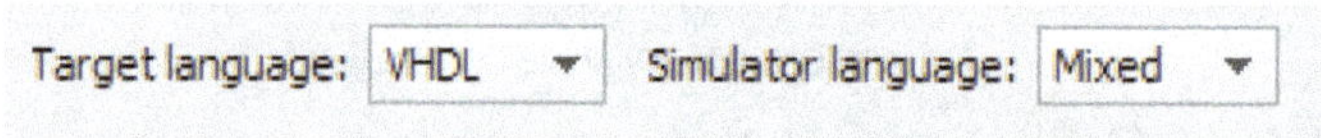

If existing sources, in the form of HDL or netlist files, were to be added to the project they could be imported at this stage.

As we do not have any sources to add to the project, click ***Next***.

(g) The *Add Existing IP (optional)* dialogue will open.

If existing IP sources were to be included in the project, they could be added here.

As we do not have any existing IP to add, click ***Next***.

(h) The *Add Constraints (optional)* dialogue will open.

This is the stage where any physical or timing constraints files could be added to the project.

As we do not have any constraints files to add, click ***Next***.

(i) From the *Default Part* dialogue,

Zed

Select ***Boards*** from the *Specify* box and choose ***ZedBoard Zynq Evaluation and Development Kit*** from the *Display Name* list and **All** from the *Board Rev* list, as shown in Figure 2.3. Select the appropriate revision for your board (in this case **Rev. D** has been selected).

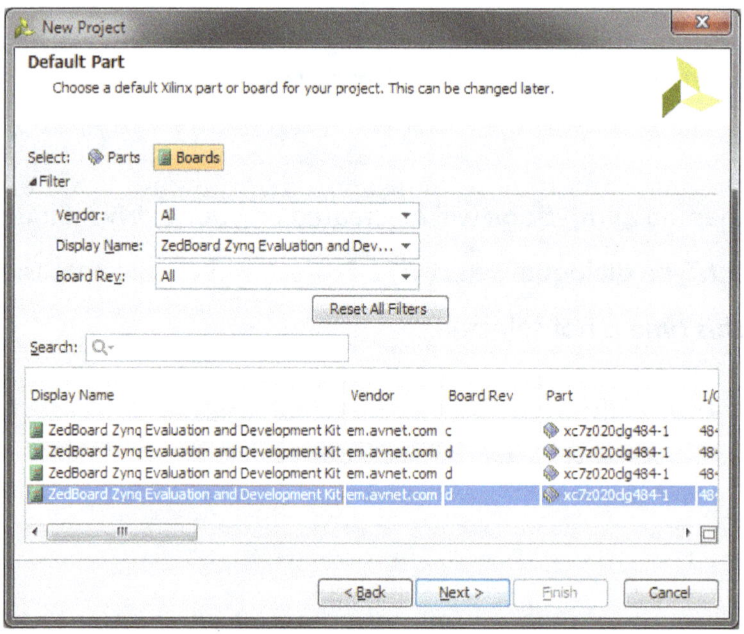

Figure 2.3: Zedboard Default Part Dialogue Options

Click ***Next***.

Zybo Ensure you have carried out the Zybo board part set-up procedure at the beginning of Exercise 1A. Select ***Boards*** from the *Select* dialogue click ***Zybo*** from the *Display Name* list and **All** from the *Board Rev* list, as shown in Figure 2.4. Select the appropriate revision for your board (in this case **Rev. B.3** has been selected).

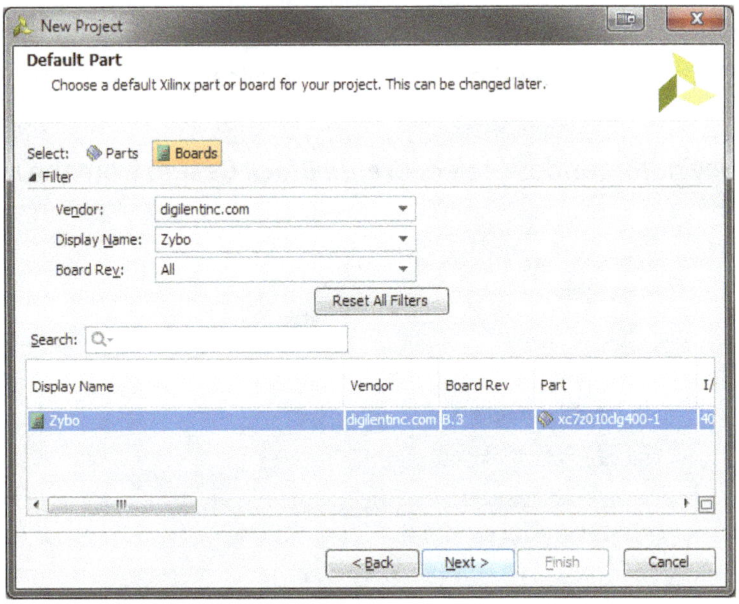

Figure 2.4: Zybo Default Part Dialogue Options

Click ***Next***.

Resume (j) In the *New Project Summary* dialogue, review the specified options, and click ***Finish*** to create the project.

As in the previous tutorial we will now create the basic Zynq embedded system design before adding and configuring additional IP to utilise hardware interrupts.

Exercise 2B Creating a Zynq System with Interrupts in Vivado

In this exercise we will create a simple Zynq embedded system which implements two General Purpose Input/Output (GPIO) controllers in the PL of the Zynq device, one of which uses the Zynq development board's push buttons to generate interrupts. The other GPIO controller will connect to the LEDs. Both will also be connected to the Zynq processor via an AXI bus connection, allowing the LEDs to be controlled by a software application which we will create in Exercise 2C.

(a) In the *Flow Navigator* window, select ***Create Block Design*** from the *IP Integrator* section, as in Figure 2.5:

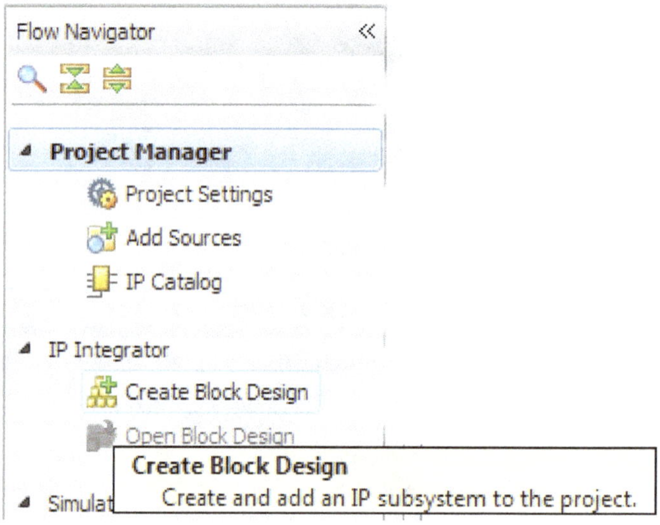

Figure 2.5: Creating a New Block Design in Flow Navigator

The *Create Block Design* dialogue will open.

(b) Enter ***zynq_interrupt_system*** in the Design name box, as in Figure 2.6:

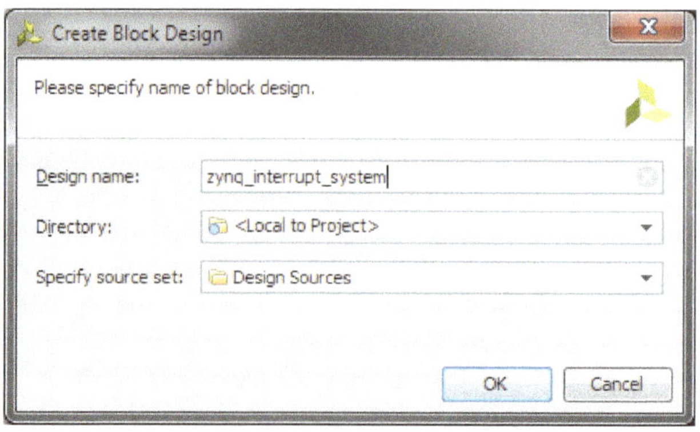

Figure 2.6: Create Block Design Dialogue

Click ***OK***. The *Vivado IP Integrator Diagram* canvas will open in the *Workspace*.

The first block that we will add to our design will be a Zynq Processing System.

(c) In the *Vivado IP Integrator Diagram* canvas, right-click anywhere and select ***Add IP***, as in Figure 2.7.

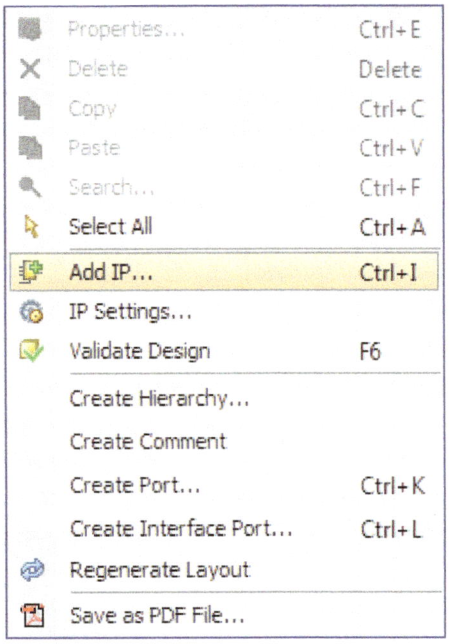

Figure 2.7: Add IP Option

Alternatively, select the ***Add IP*** button in the toolbar at the left of the canvas, shown in Figure 2.8.

Figure 2.8: Add IP option in IP Integrator canvas information message

The pop-up IP Catalog window will open, as in Figure 2.9.

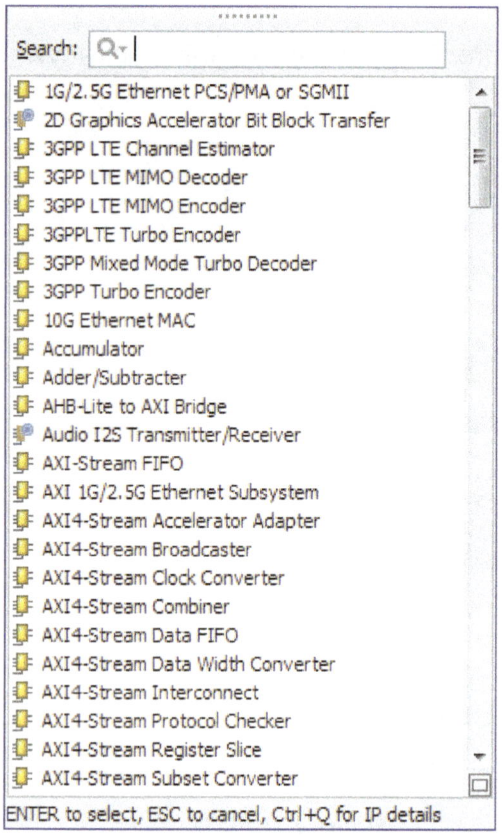

Figure 2.9: Pop-up IP Catalog Window

Enter ***zynq*** in the search field and select the ***ZYNQ7 Processing System***, as shown in Figure 2.10, and press the ***Enter*** key on your keyboard.

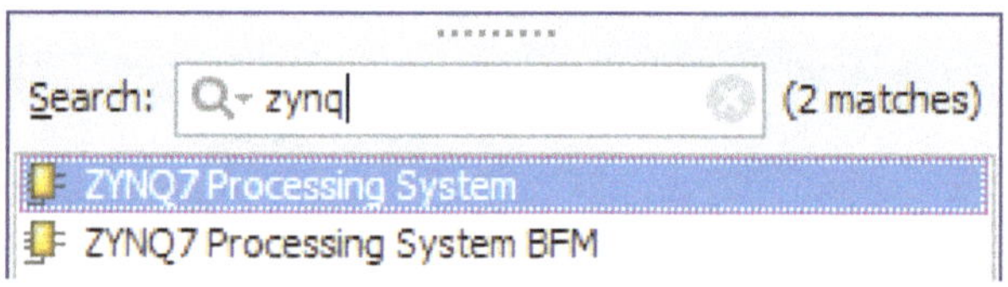

Figure 2.10: Adding ZYNQ7 Processing System from IP Catalog

(d) As in the previous tutorial, the next step is to connect the **DDR** and **FIXED_IO** interface ports on the Zynq PS to the top-level interface ports on the design.

Zed

Select the ***Run Block Automation*** option from the *Designer Assistance* message at the top of the Diagram window. Select ***OK***, ensuring that the option to ***Apply Board Preset*** is selected, to generate the external connections for both the **DDR** and **FIXED_IO** interfaces, and apply the relevant board presets.

Your block diagram should now resemble Figure 2.11.

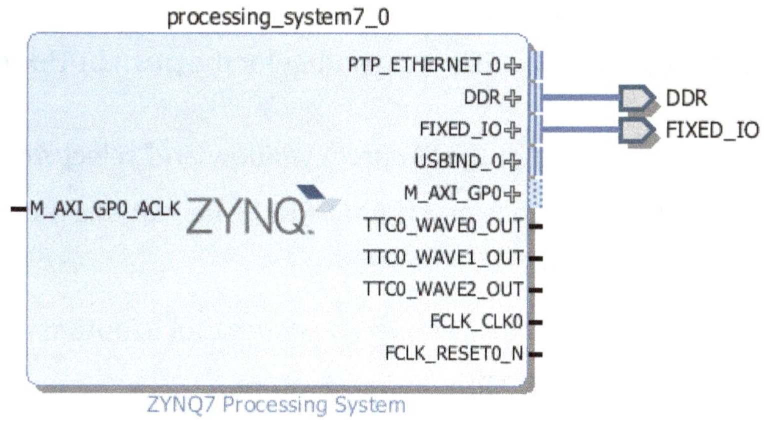

Figure 2.11: Zedboard ZYNQ7 Processing System External Connections

As the **Zedboard** platform is the target development board, and this was specified on creation of the project, Vivado will configure the Zynq processor block accordingly.

Zybo

In the *Run Block Automation* dialogue, ensure that the option to ***Apply Board Preset*** is selected and click ***OK***. The external connections for both the **DDR** and **FIXED_IO** interfaces will now be generated.

Your block diagram should now resemble Figure 2.12.

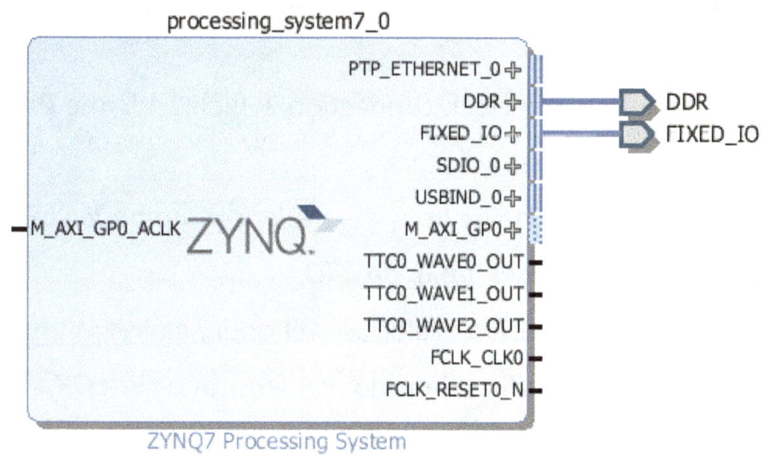

Figure 2.12: Zybo ZYNQ7 Processing System External Connections

As the ***Zybo*** platform is the target development board, and this was specified on creation of the project, Vivado will configure the Zynq processor block accordingly.

Resume

Now that the main Zynq PS has been added to our design and configured, we can now add further blocks which will be placed in the PL to add functionality to the system. In this case we require an **AXI GPIO** block for the **LEDs** and another for the **push buttons**.

(e) Right-click in an empty area of the *Diagram* window and select ***Add IP***. Enter ***GPIO*** in the search field and add an instance of the ***AXI GPIO*** IP. Repeat this procedure to add a second **AXI GPIO** block to the design.

(f) We will now use the *IP Integrator Designer Assistance* tool to automate the connection of the **AXI GPIO** blocks to the **ZYNQ7 Processing System**.

Zed

Click ***Run Connection Automation*** from the *Designer Assistance* message at the top of the *Diagram* window and select ***/axi_gpio_0/S_AXI***, as shown Figure 2.13.

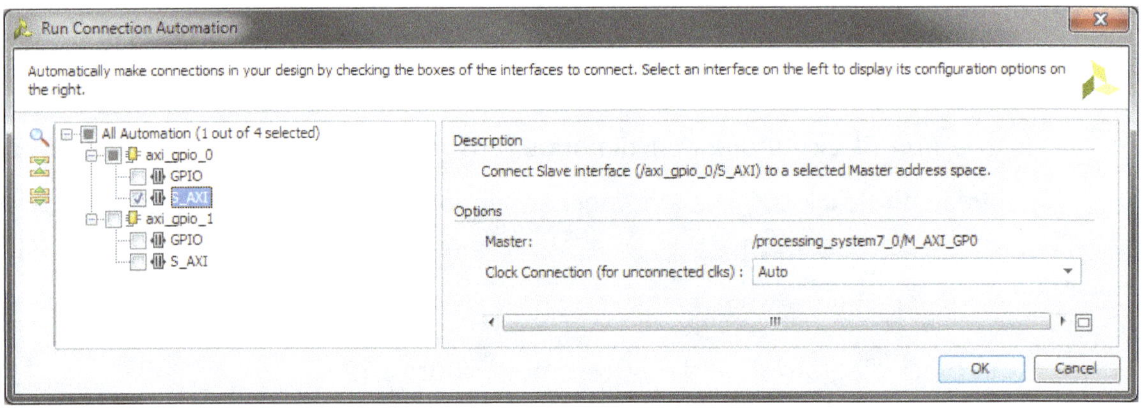

Figure 2.13: Zedboard Run Block Automation - GPIOinstance

Click ***OK*** to ensure automatic clock connection, which adds the **Processor System Reset Module** and the **AXI Interconnect** blocks.

Click ***Run Connection Automation*** from the *Designer Automation* message at the top of the *Diagram* window and select ***/axi_gpio_0/GPIO***.

The ***Run Connection Automation*** dialogue will open, as in Figure 2.14. Select ***btns_5bits*** from the *Select Board Part Interface* drop-down menu, and click ***OK***.

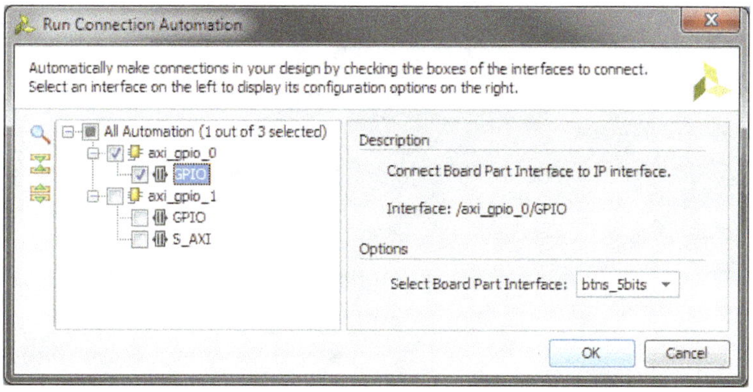

Figure 2.14: Zedboard Run Connection Automation dialogue — GPIO

Repeat step **(f)** for the second **GPIO** block, this time selecting ***leds_8bits*** for ***/axi_gpio_1/ GPIO.*** This will result in a system that is similar to Figure 2.15.

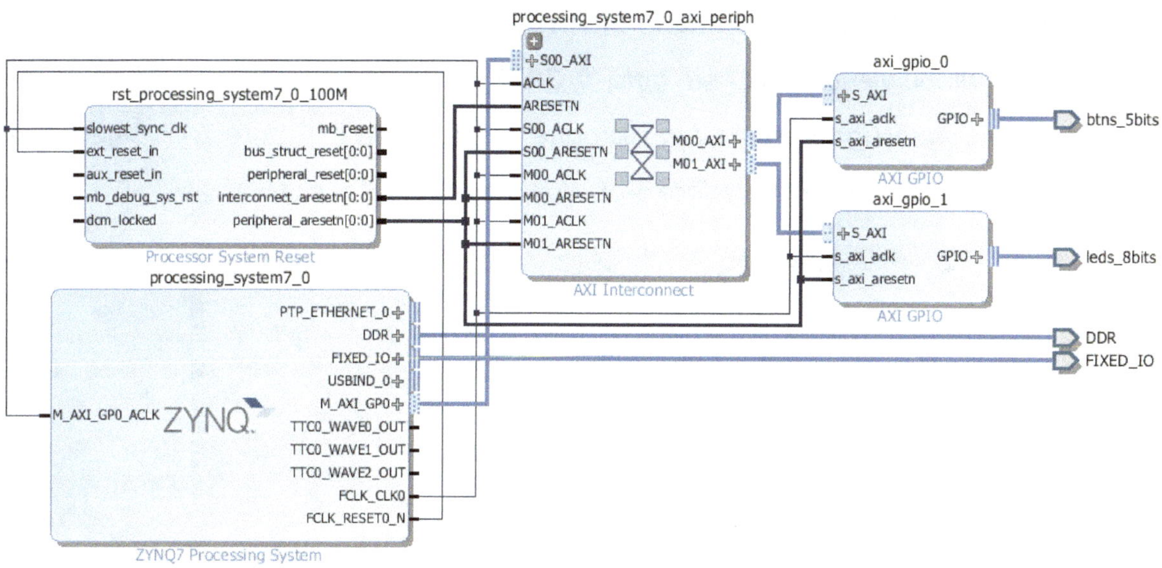

Figure 2.15: Zedboard Zynq Processor System

Zybo Click ***Run Connection Automation*** from the *Designer Assistance* message at the top of the *Diagram* window and select ***/axi_gpio_0/S_AXI***, as shown Figure 2.16.

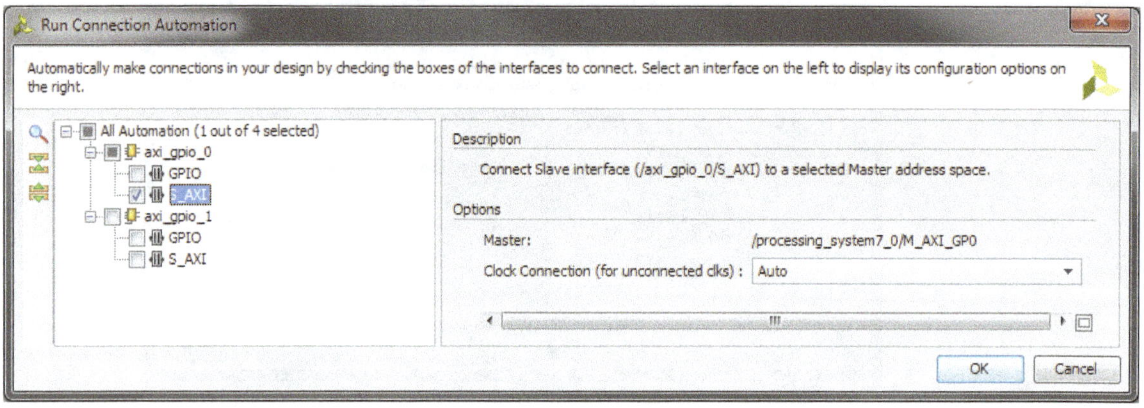

Figure 2.16: Zybo Run Block Automation - GPIOinstance

Click ***OK*** to ensure automatic clock connection, which adds the **Processor System Reset Module** and the **AXI Interconnect** blocks.

Click ***Run Connection Automation*** from the *Designer Automation* message at the top of the *Diagram* window and select ***/axi_gpio_0/GPIO.***

The ***Run Connection Automation*** dialogue will open, as in Figure 2.17. Select ***btns_4bits*** from the *Select Board Part Interface* drop-down menu, and click ***OK.***

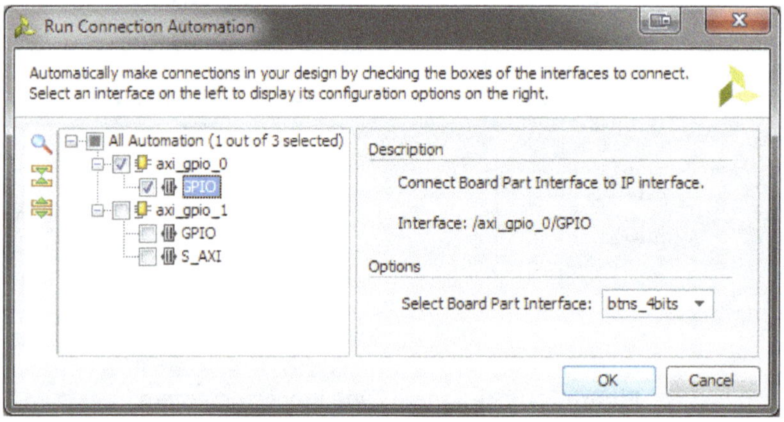

Figure 2.17: Zybo Run Connection Automation dialogue — GPIO

Repeat step **(f)** for the second **GPIO** block, this time selecting ***leds_4bits*** for ***/axi_gpio_1/ GPIO.*** This will result in a system that is similar to Figure 2.18.

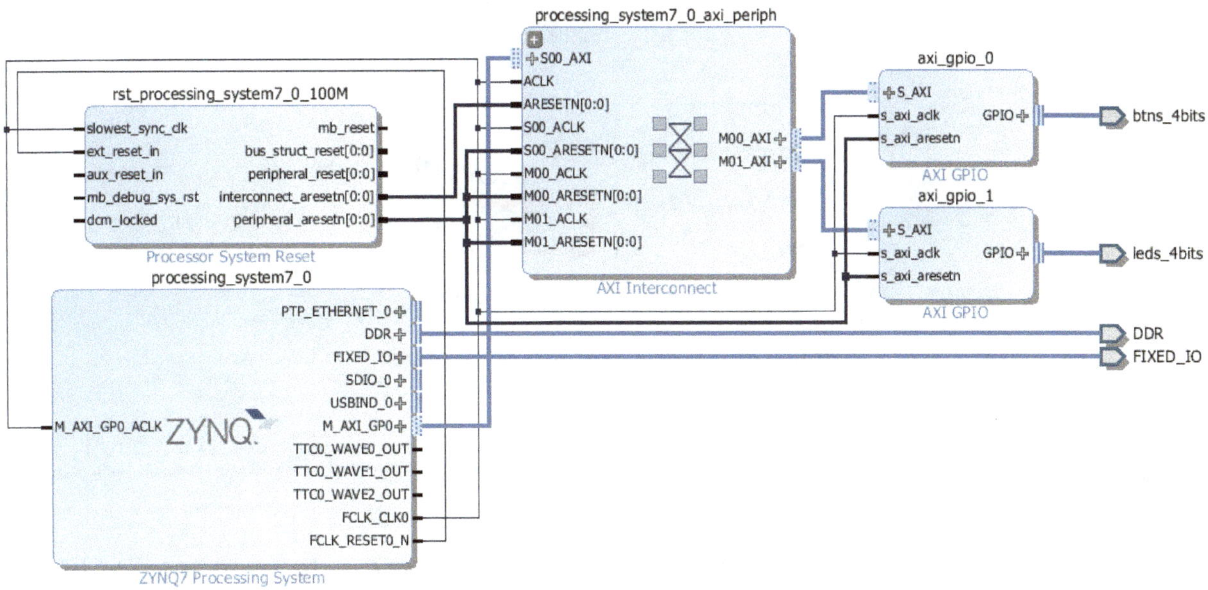

Figure 2.18: Zybo Zynq Processor System

Resume We now need to configure the system to utilise hardware interrupts from the **push buttons** to trigger functions in the **Zynq PS**. Return to the *Block Diagram*.

(g) Double-click on the **axi_gpio_0** block, which is connected to the **push buttons**, to open the *Re-customize IP* window. Select **IP Configuration** tab as shown below:

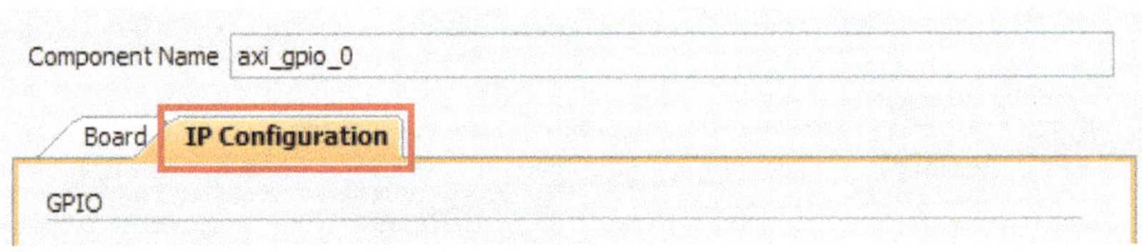

In the **IP Configuration** window, enable interrupts from the **push buttons** by clicking in the box highlighted in Figure 2.19 and click ***OK.***

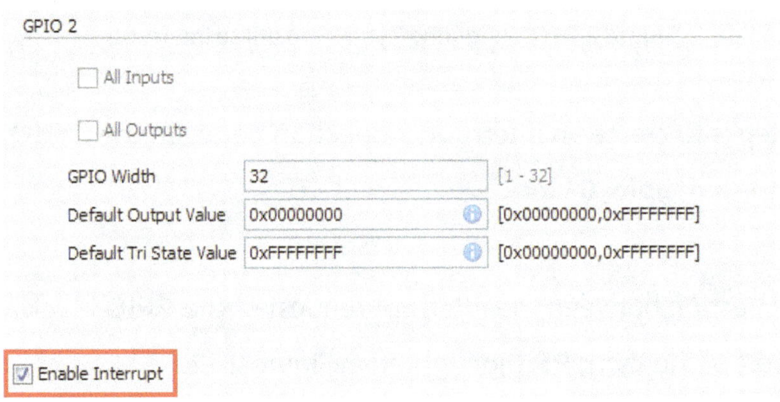

Figure 2.19: Enabling GPIO interrupts

This will add an additional output port for the interrupt request to the **GPIO** block as shown in Figure 2.20.

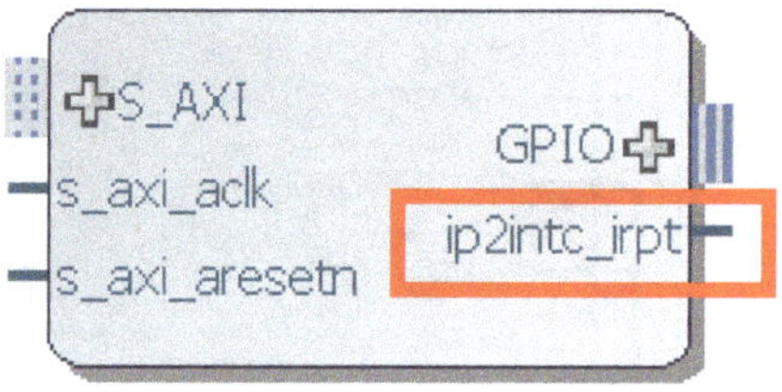

Figure 2.20: GPIO block with interrupt port

Now we must configure the **Zynq PS** to accept interrupt requests.

(h) Double-click on the **Zynq PS** block, ***processing_system7_0***, to open the *Re-Customize IP* window.

(i) Select *Interrupts* from the *Page Navigator* on the left-hand side and expand the menu on the right as in Figure 2.21. Since we want to allow interrupts from the **programmable logic** to the **processing system,** tick the box to enable **Fabric Interrupts**, then click to enable the shared interrupt port as in Figure 2.21. This means interrupts from the **PL** can be connected to the interrupt controller within the **Zynq PS**. Click ***OK.***

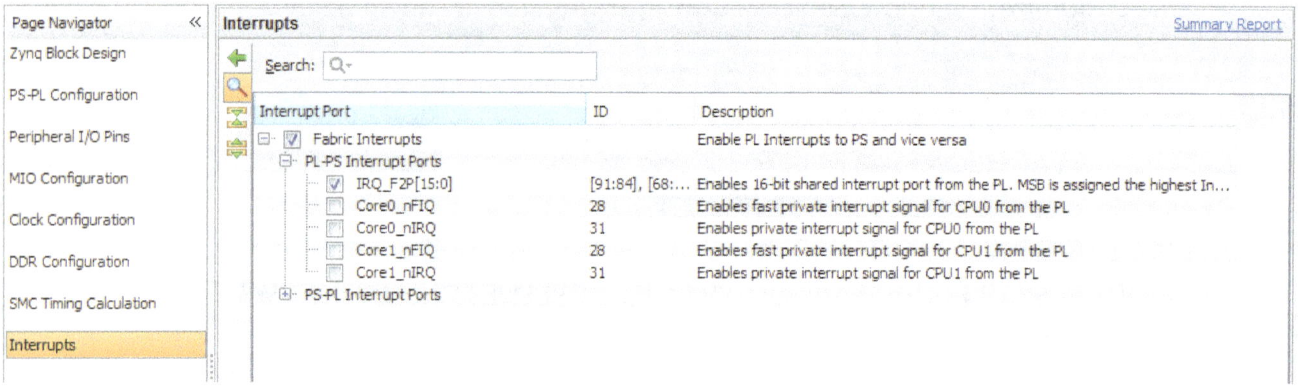

Figure 2.21: Configuring Zynq PS to utilise interrupts

(j) The final step is to create an interrupt connection between the **ZYNQ7 Processing System** block and the **axi_gpio_0** block.

Zed Make a connection between the interrupt request of the **GPIO** block and the newly created interrupt port of the **Zynq PS**, highlighted in Figure 2.22.

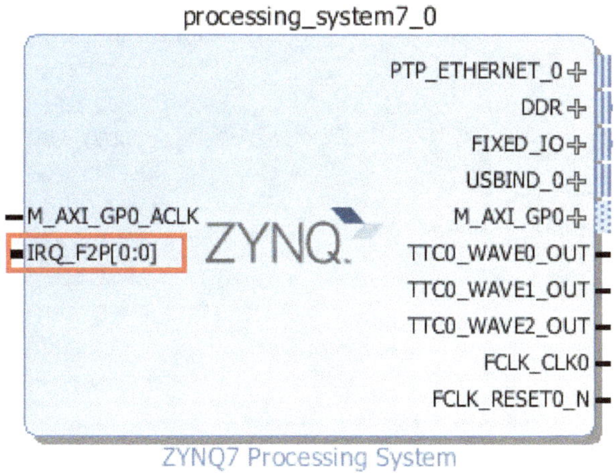

Figure 2.22: Zedboard Zynq PS with interrupt port

Your final design should resemble Figure 2.23, although the positioning of your blocks may be different.

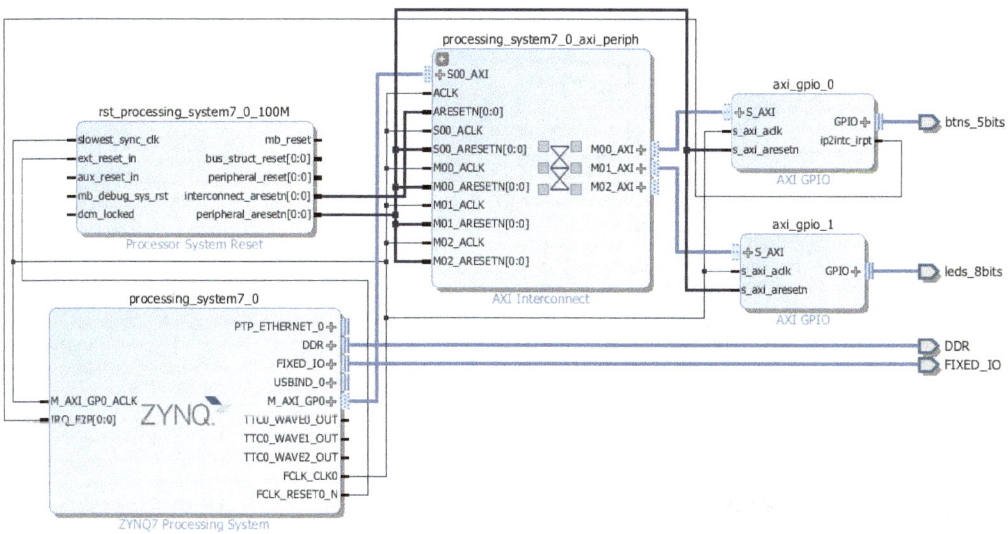

Figure 2.23: Zedboard Zynq processor system with interrupts

Zybo Make a connection between the interrupt request of the **GPIO** block and the newly created interrupt port of the **Zynq PS**, highlighted in Figure 2.24.

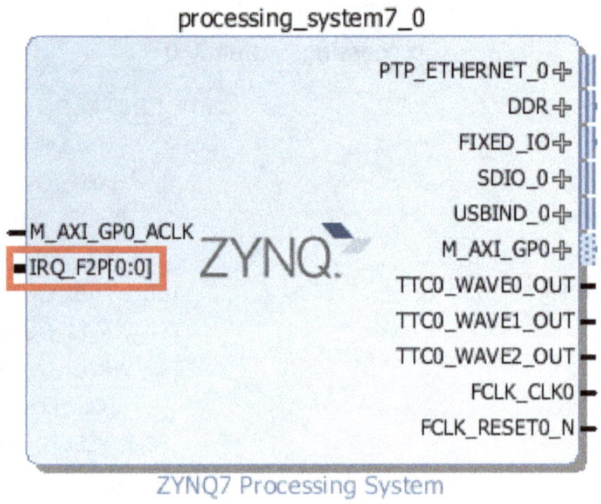

Figure 2.24: Zybo Zynq PS with interrupt port

Your final design should resemble Figure 2.25, although the positioning of your blocks may be different.

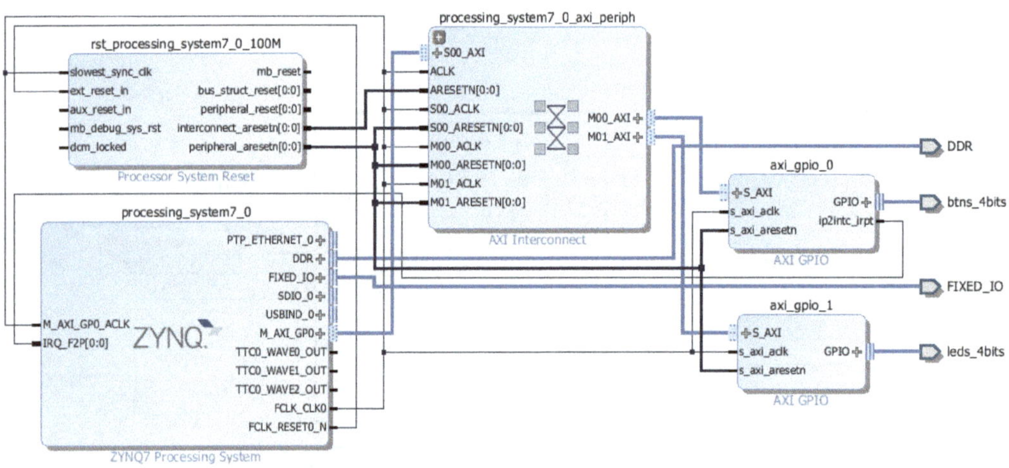

Figure 2.25: Zybo Zynq processor system with interrupts

Resume (k) Save your design by selecting ***File > Save Block Design*** from the *Menu Bar.*

(l) Validate the design by selecting ***Tools > Validate Design*** from the *Menu Bar*. This will run a Design-Rule-Check (DRC).

Alternatively, select the Validate Design button, , from the Main Toolbar.

(m) A *Validate Design* dialogue should appear to confirm that validation of the design was successful. Click ***OK,*** to dismiss the message.

With the design successfully validated, we can now move on to generating the HDL design files for the system. The procedure here is identical to the previous tutorial, ***First Designs on Zynq***.

(n) In the **Sources** window of the *Data Windows* pane, select the ***Sources*** tab.

(o) Right-click on the top-level system design, which in this case is **zynq_interrupt_system**, and select ***Create HDL Wrapper***, as shown in Figure 2.26.

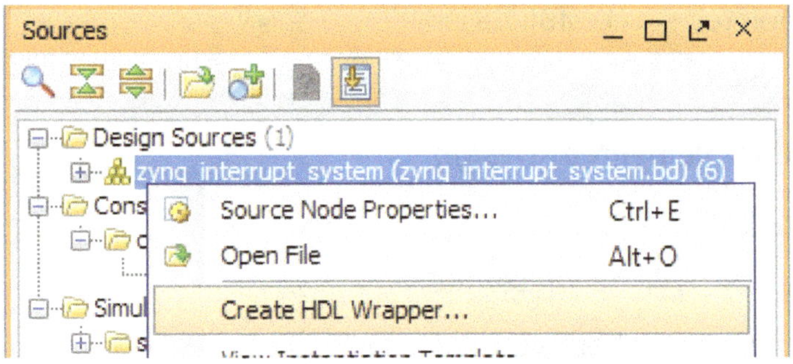

Figure 2.26: Create HDL Wrapper

The *Create HDL Wrapper* dialogue window will open. Accept the default option specifying that Vivado should manage the wrapper and click ***OK***.

With all HDL design files generated, the next step in Vivado is to implement our design and generate a bitstream file.

(p) In *Flow Navigator*, click ***Generate Bitstream*** from the *Program and Debug* section. If a dialogue window appears prompting you to save your design, click ***Save***.

(q) A dialogue window will open requesting that you launch synthesis and implementation before starting the *Generate Bitstream* process. Click ***Yes*** to accept.

The combination of running the synthesis, implementation and bitstream generation processes back-to-back may take a few minutes, depending on the power of your computer system.

(r) Once the bitstream generation is complete a dialogue window will open to inform you that the process has been completed successfully, as in Figure 2.27.

Select ***Open Implemented Design***, and click ***OK***.

At this point you will be presented with the *Device* view, where you can see the PL resources which are utilised by the design.

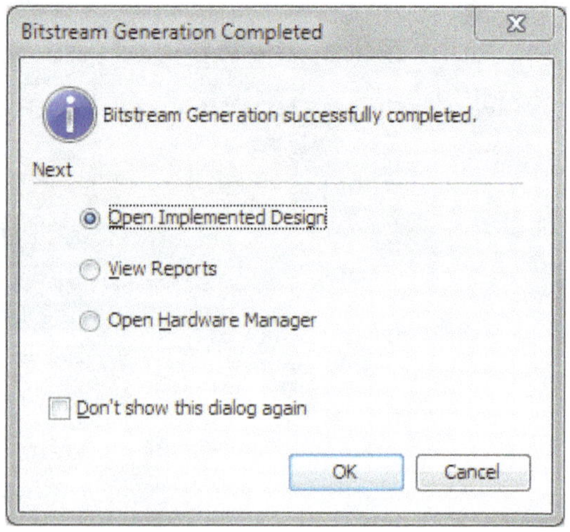

Figure 2.27: Bitstream Generation completion dialogue

With the bitstream generation complete, the final step in Vivado is to export the design to the SDK, where we will create the software application that will allow the Zynq PS to control the LEDs on the development board.

(s) Select ***File > Export > Export Hardware*** from the *Menu Bar*.

(t) The *Export Hardware for SDK* dialogue window will open. Ensure that the option to ***Include bitstream*** is selected, as in Figure 2.28, and Click ***OK***.

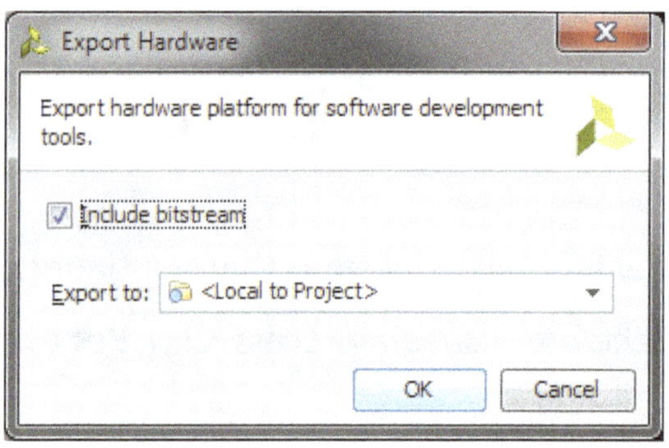

Figure 2.28: Export Hardware for SDK

(u) Launch the SDK in Vivado by selecting ***File > Launch SDK*** from the *Menu Bar* and Click ***OK***.

This concludes the steps that are required in Vivado IDE. All hardware components of the system have been configured and generated. In the next exercise we will create the software application that utilises this hardware system.

Exercise 2C Creating a Software Application in the SDK

In this exercise a software application will be created that utilises hardware interrupts on the Zynq development board. The push buttons will be used to increment a counter by different values, and the count will be continuously displayed on the LEDs in binary form, where LED0 corresponds to the least significant bit (LSB) and the uppermost LED is the most significant bit (MSB). This application will run on the Zynq processing system, communicating with the AXI GPIO blocks implemented in the PL.

The SDK should have opened after the conclusion of Exercise 2B. If it did not open, you can open the SDK by navigating to **Start > All Programs > Xilinx Design Tools >SDK 2015.1> Xilinx SDK 2015.1** and specifying the workspace as in Exercise 1C.

(a) Select **File > New > Application Project** from the *Menu bar*.

(b) The *New Project* dialogue window will open. Enter ***interrupt_counter*** in the *Project name* field, as shown in Figure 2.29, keeping all other options with the default settings. Click ***Next***.

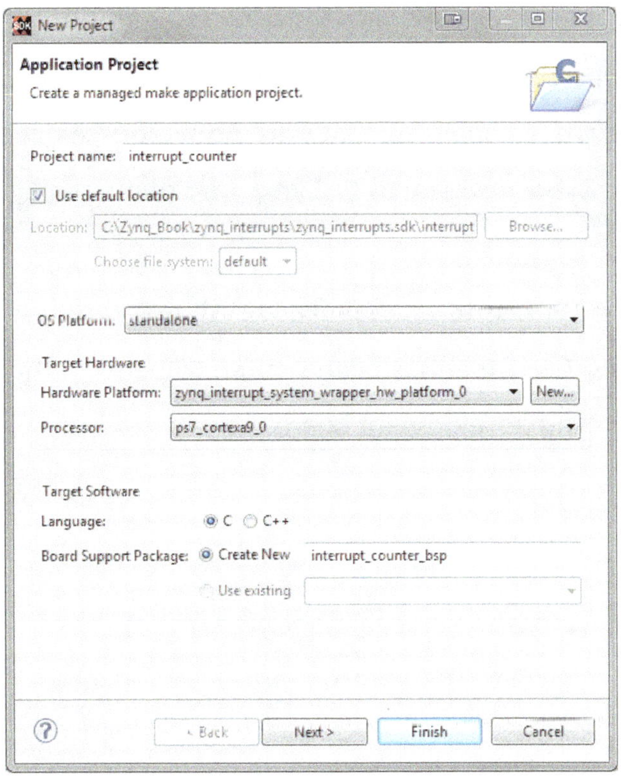

Figure 2.29: New Application Project dialogue

(c) At the *New Project Templates* screen, select ***Empty Application***, as in Figure 2.30, and click ***Finish*** to create the project.

NOTE: The new project should open automatically. If it doesn't, you may need to close the welcome tab in order to view the project.

With the new Application Project created, we can now import pre-prepared source code for the application.

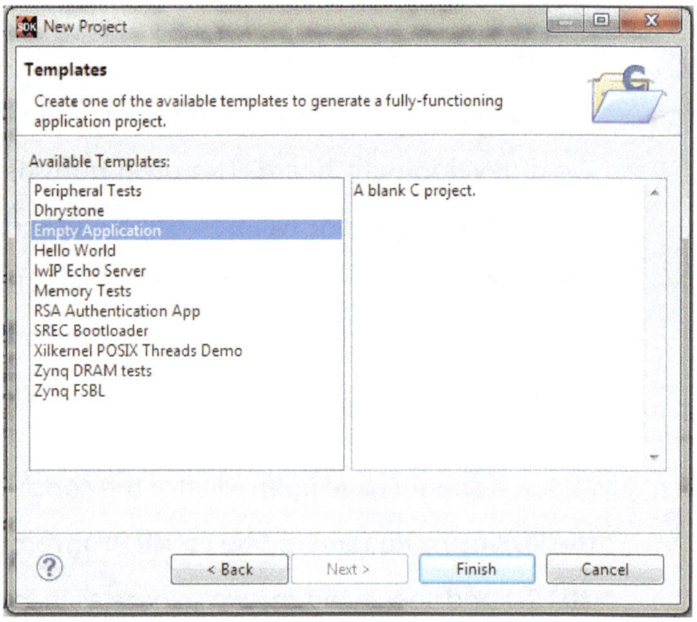

Figure 2.30: New Project Template dialogue

(d) In the *Project Explorer* panel, expand ***interrupt_counter*** and highlight the *src* directory. Right-click and select ***Import...***, choosing ***General*** > ***File System*** as an import source.

(e) In the *Import File System* window, click the ***Browse...*** button.

(f) The source file directory will depend on the Zynq development board that is in use. If you are using the **Zedboard**, navigate to: **C:\Zynq_Book\sources\zedboard\zynq_interrupts.** If you are using the **Zybo**, navigate to **C:\Zynq_Book\sources\zybo\zynq_interrupts.**
Click ***OK.***

(g) Select the file **interrupt_counter_tut_2B.c**, as shown in Figure 2.31, and click ***Finish***.

This file contains C Code that has been written to perform the interrupt triggered counter operation on the Zynq development board.

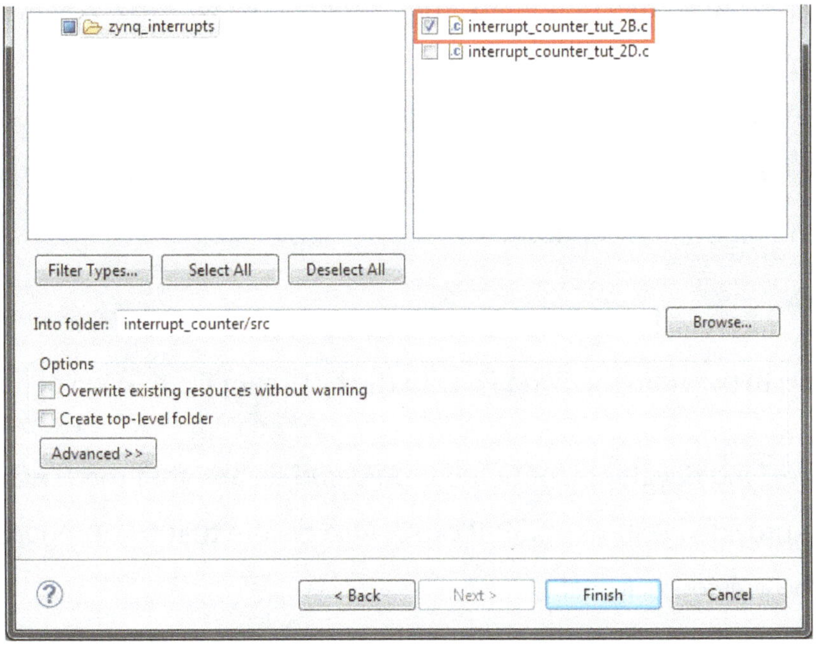

Figure 2.31: Import C source file

(h) Open the imported source file by expanding the *src* folder and double-clicking on **interrupt_counter_tut_2B.c**, and explore the code.

The code has been fully commented, but will be briefly discussed here for clarity.

By now, you should be familiar with the use of drivers and parameters in configuring and operating the GPIO. Remember, detailed information of these drivers can be found in the **system.mss** file, explaining the purpose of each function and the parameters passed to it. Predesignated parameters can also be found in `xparameters.h`.

The Zynq PS features a built in interrupt controller, initialised here as `XScuGic INTCInst`. This handles all incoming interrupt requests passed to the PS and performs the function associated with each interrupt source. It is also capable of prioritising multiple interrupt sources to the requirements of the application.

Of particular note is the inclusion of the function `BTN_Intr_Handler(void *InstancePtr);`. This is the interrupt handler function for the push buttons and is called every time an interrupt request from the push buttons in the PL is received in the PS. This performs a counter increment on each call and displays the value of the counter on the LEDs in binary.

An initial setup function can be found below the main function. This is `InterruptSystemSetup(XScuGic *XScuGicInstancePtr);`. The function initialises and configures the interrupt controller in the Zynq PS, connecting the interrupt handler to the interrupt source. It also makes a call to the latter function which enables the interrupt sources and registers exceptions.

The next step is to program the Zynq PL with the bitstream file that we generated in Exercise 2B.

Ensure that the Zynq development board is powered on and that the JTAG port is connected to the PC via the provided USB-A to USB-B cable. Also ensure that the jumper positions are correct as shown in the previous tutorial.

(i) Download the bitstream to the Zynq PL by selecting **Xilinx Tools > Program FPGA** from the *Menu bar*. The *Program FPGA* window will appear. The Bitstream field should already be populated with the correct bitstream file, as in Figure 2.32.

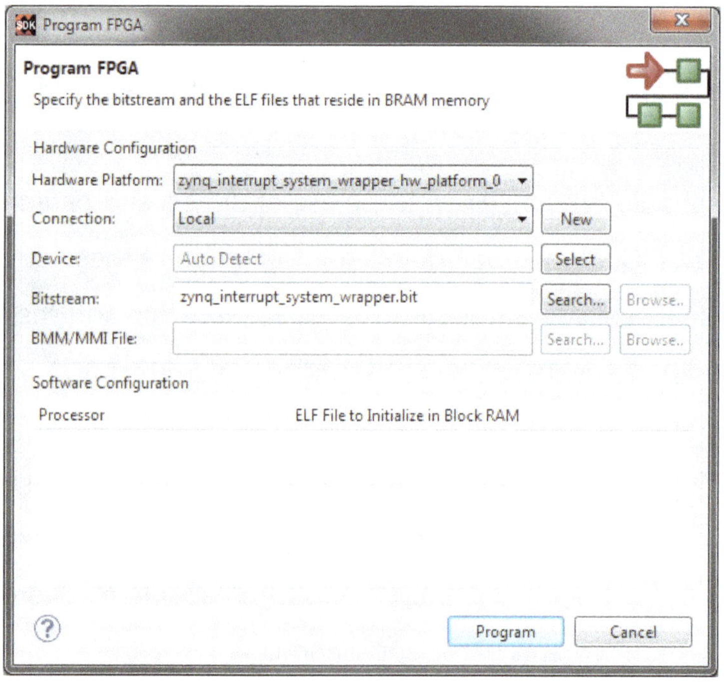

Figure 2.32: Program FPGA dialogue window

As in the previous tutorial, once the device has successfully been programmed, the *DONE LED* on the ***ZedBoard*** will turn blue. Similarly the *DONE LED* on the ***Zybo*** will turn green.

With the Zynq PL successfully configured with the bitstream file, we can now launch our software application on the Zynq PS.

(j) Select ***interrupt_counter*** in *Project Explorer*. Right-click and select **Run As > Launch on Hardware (GDB)**.

Zed

The counter increments by different values based on the push buttons which are pressed. This operation is demonstrated in Figure 2.33.

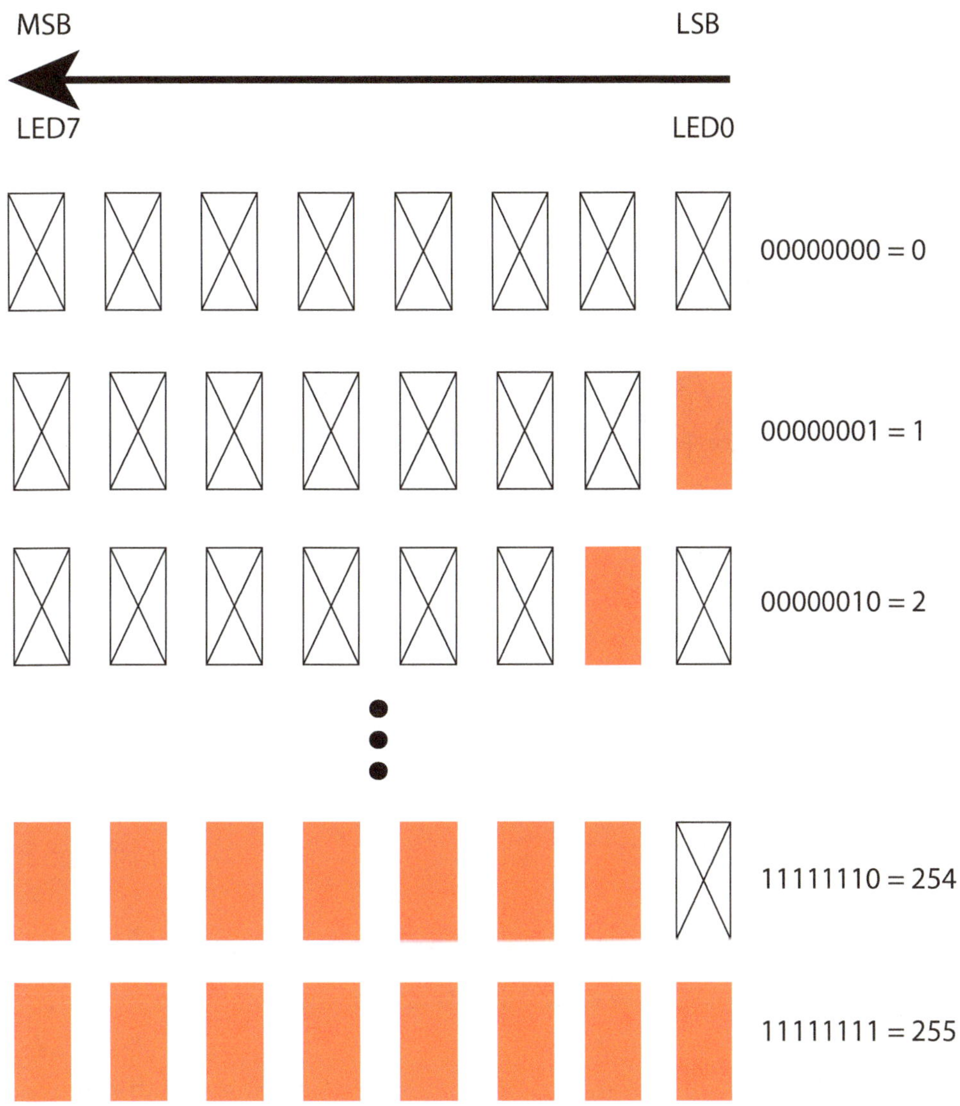

Figure 2.33: Zedboard LED flashing states

Try pressing different push buttons and observing how the counter increments (can the counter achieve 255?) Based on your findings, can you determine the value assigned to each of the push buttons (BTNU, BTND, BTNL, BTNR and BTNC as noted on the ZedBoard)?

Zybo

The counter increments by different values based on the push buttons which are pressed. This operation is demonstrated in Figure 2.34.

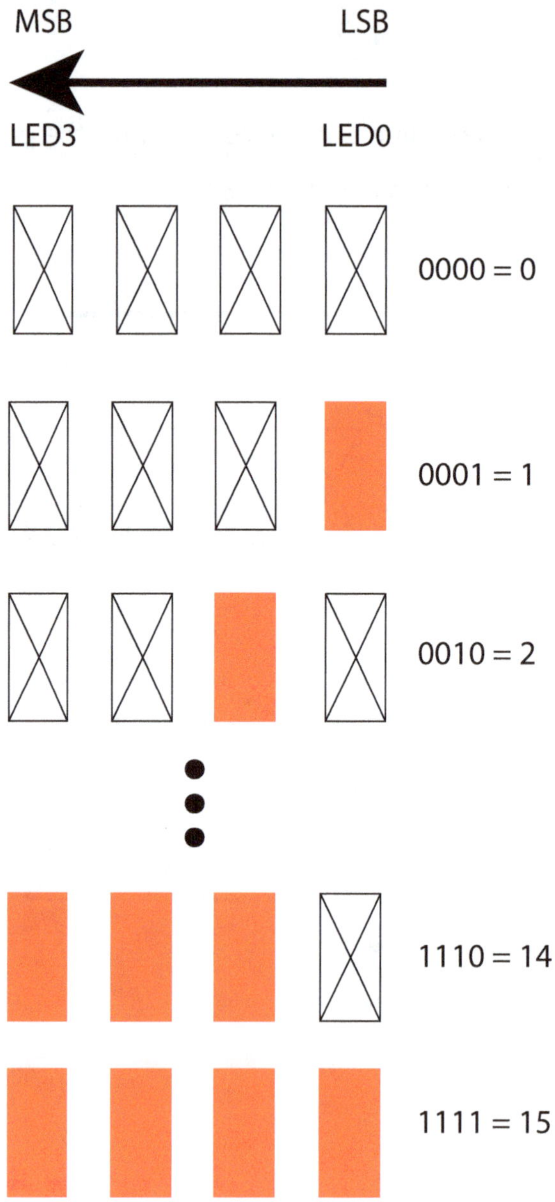

Figure 2.34: Zybo LED flashing states

Try pressing different push buttons and observing how the counter increments. Based on your findings, can you determine the value assigned to each of the push buttons (BTN0, BTN1, BTN2, BTN3 as noted on the Zybo)?

Resume You have successfully created and executed a software application utilising interrupts on the Zynq PS. The next step is to go back and add an additional interrupt source with higher priority to alter the functionality of the system.

Exercise 2D Adding a Further Interrupt Source

In this exercise we will add an additional source of interrupt to the project created in Exercise 2B in the form of an **AXI Timer**.

(a) Launch Vivado by double-clicking on the Vivado desktop icon: , or by navigating to **Start > All Programs > Xilinx Design Tools > Vivado 2015.1 > Vivado 2015.1**

(b) When the program launches, open the previously created project by selecting ***Open Project***. The previously created project should appear in the list of recent projects as ***C:/Zynq_Book/zynq_interrupts/zynq_interrupts.xpr*** so click on it. If it doesn't, click ***Browse Projects...*** and navigate to that directory, selecting ***zynq_interrupts.xpr*** and clicking open.

(c) Open the block design from the sources panel by expanding the sources and double clicking on the block design as highlighted in Figure 2.35.

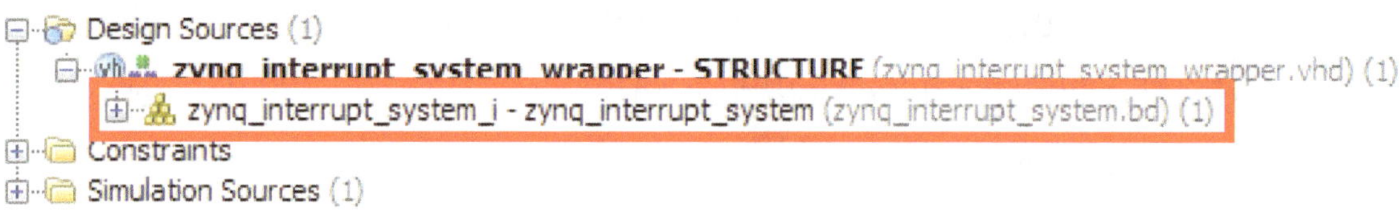

Figure 2.35: Opening an existing block diagram

(d) With the block diagram now open we will add an **AXI Timer** to the design. In the *Vivado IP Integrator Diagram* canvas, right-click anywhere and select ***Add IP***. Enter ***timer*** in the search field and add the IP ***AXI TIMER*** to the design by either dragging it onto the canvas or selecting it and pressing ENTER.

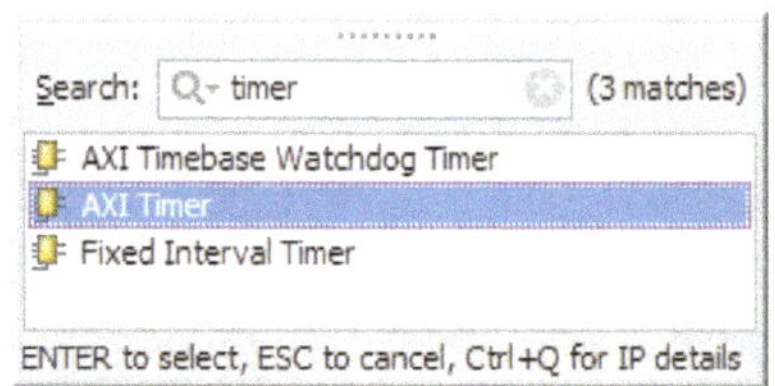

Figure 2.36: AXI Timer in the IP Catalog

(e) Select ***Run Connection Automation*** option from the *Designer Assistance* message at the top of the Diagram window. In the ***Run Connection Automation*** window, select**/axi_timer_0/S_AXI** to connect the timer to the AXI Interconnect. Click ***OK***..

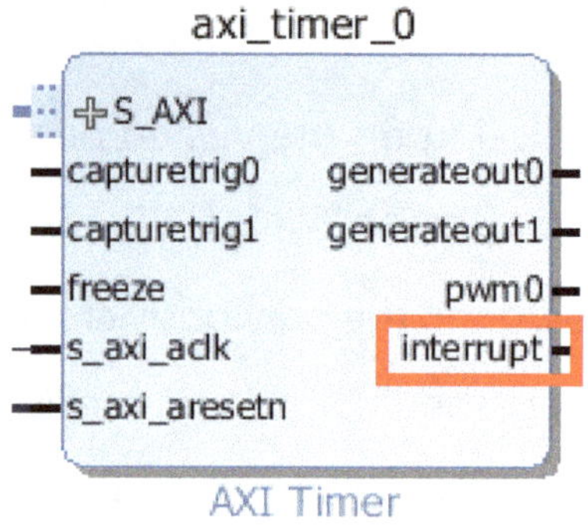

Figure 2.37: AXI Timer in the block design

(f) Note that in Figure 2.37 the AXI Timer features an interrupt request, which requires connection to the Zynq PS. However, we already have an interrupt connected to the input of the PS. This input is a shared interrupt port, and so accepts multiple interrupts via one signal. We therefore require an additional IP block to concatenate these two interrupt requests into one signal. In the canvas, right-click anywhere and select ***Add IP***. Enter ***concat*** in the search field and add the IP ***Concat*** to the design.

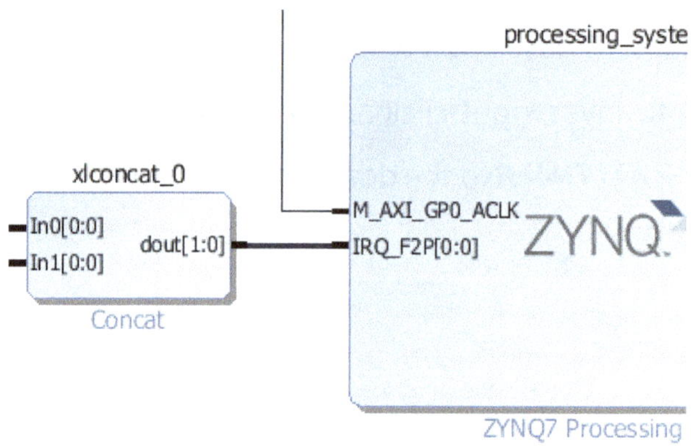

Figure 2.38: Concat in the block design

(g) Remove the connection between the **AXI_GPIO ip2intc_irpt** and **IRQ_F2P[0:0]** on the Zynq PS by clicking on the line between and pressing DELETE. Connect the output from he ***Concat*** block, ***xlconcat_0*** to this instead. Then, connect the interrupt request from the GPIO to **In0[0:0]** and the interrupt from the timer to **In1[0:0]**, creating a shared interrupt signal that is passed to the PS.

Zed Your block diagram should be similar to Figure 2.39.

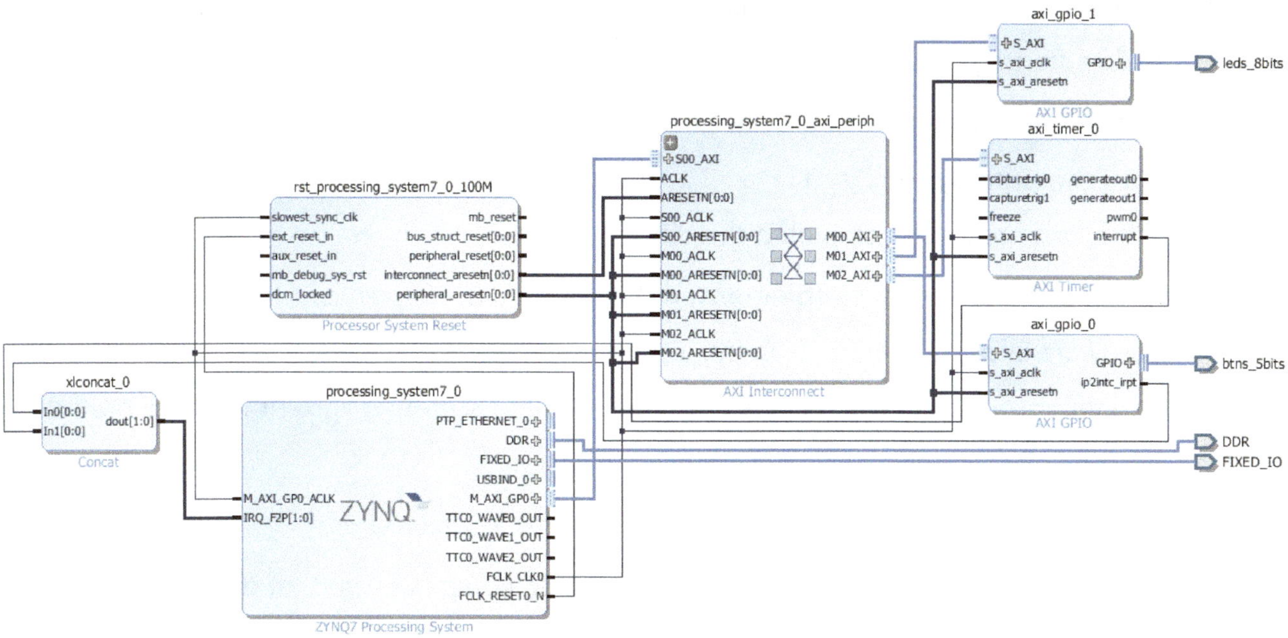

Figure 2.39: Zedboard complete system with multiple interrupts sources

Zybo Your block diagram should be similar to Figure 2.40.

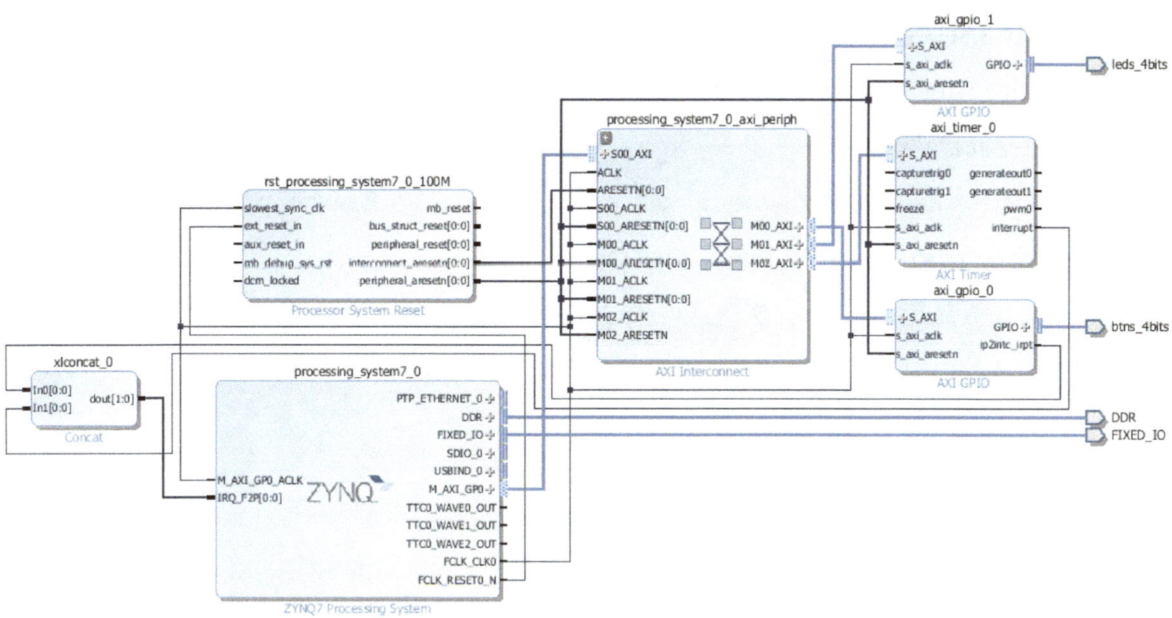

Figure 2.40: Zybo complete system with multiple interrupts sources

Resume We now need to generate a new bitstream for our altered design.

(h) In *Flow Navigator*, click ***Generate Bitstream*** from the *Program and Debug* section.

If a dialogue window appears prompting you to save your design, click ***Save***.

(i) A dialogue window will open requesting that you launch synthesis and implementation before starting the *Generate Bitstream* process. Click ***Yes*** to accept.

Again these back-to-back processes may take a few minutes, depending on the power of your computer system.

(j) When this process is completed click ***OK***.

(k) In *Flow Navigator*, select **Implemented Design** from the *Implementation* section to open the hardware implementation diagram. If the design needs to *reload* click ***reload*** on the yellow banner at the top of the hardware implementation screen. A current implemented design must be open in order to export a hardware design.

(l) Select ***File > Export > Export Hardware...*** from the *Menu Bar*.

(m) The *Export Hardware for SDK* dialogue window will open. Ensure that the option to ***Include bitstream*** is selected, and Click ***OK***. A dialog will be presented asking if you wish to overwrite an exported file, which is the initial system featuring a single interrupt. Select ***Yes*** for this and any further prompts.

(n) Launch the SDK in Vivado by selecting ***File > Launch SDK*** from the *Menu Bar* and Click ***OK***.

Once the SDK opens and builds the project, we will alter our application to make use of the new interrupt source. Right-click on the project **interrupt_counter** in the ***Project Explorer*** and select ***Delete***.

The *Delete Resources* dialogue will open. Select the **Delete project contents on disk** checkbox as in Figure 2.41, and click **OK.**

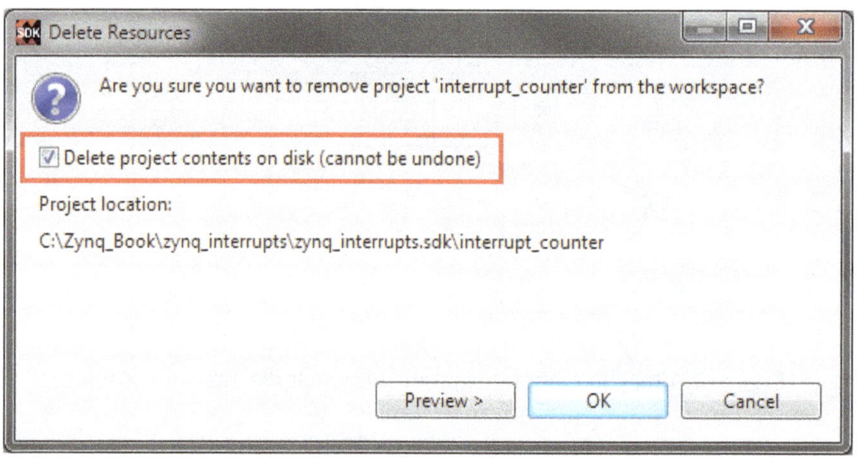

Figure 2.41: Confirm Delete Resources Dialogue

Repeat for the BSP, **interrupt_counter_bsp**. Select ***Continue*** if there are any further prompts.

Repeat the steps outlined in Exercise 2C (a) to (h) for creating a new application project, BSP and importing a source file, this time selecting **interrupt_counter_tut_2D.c**.

Notice the inclusion of a second interrupt handler, `TMR_Intr_Handler(void *data);` which will increment the value of the counter after the timer has expired three times, writing the new value to the LEDs.

Additional code has been included in the main to configure and start the timer, and full details of these functions can be found in the **system.mms**. The function `IntcInitFunction(u16 DeviceId, XTmrCtr *TmrInstancePtr,XGpio *GpioInstancePtr);` also contains additional code to connect the timer interrupt to the handler and enable it.

In brief, the timer is loaded with a value `TMR_LOAD` and configured to automatically reload on each expiration. The interrupt handler keeps track of the number of expirations and after three expirations performs the required steps, otherwise it simply increments the variable storing the number of expirations.

(o) Download the bitstream to the Zynq PL by selecting **Xilinx Tools > Program FPGA** from the *Menu bar*.

(p) Once the development board LED indicating successful programming lights up, select ***interrupt_counter*** in *Project Explorer*. Right-click and select **Run As > Launch on Hardware (GDB)**.

Note that the counter will increment by 1 when timer expires three times. The buttons still operate as in the previous exercise.

This completes this tutorial where systems utilising both single and multiple interrupt sources have been created and tested.

The Zyng Book Tutorials 3

Designing With Vivado HLS

v1.4, June 2015

Revision History

Date	Version	Changes
30/10/2013	1.0	First release for Vivado Design Suite version 2013.2
28/01/2014	1.1	Updated for changes in Vivado Design Suite version 2013.4
06/5/2014	1.2	Updated for changes in Vivado Design Suite version 2014.1
09/04/2015	1.3	Updated for changes in Vivado Design Suite Version 2014.4
30/04/2015	1.3.1	Updated to include Zybo development board for Vivado Design Suite version 2014.4
19/06/2015	1.4	Updated for changes in Vivado Design Suite Version 2015.1

Introduction

This tutorial presents an introduction to High Level Synthesis using the Vivado™ HLS environment. The creation of projects manually through the GUI, and automatically through scripting will be covered. The process of simulating, synthesising and analysing a Vivado HLS design will then be explored, with design optimisation and solution comparison along the way.

The tutorial is split into three exercises, and is organised as follows:

Exercise 3A — This exercise concerns the creation of projects using both the Vivado HLS GUI and use of Tcl scripting. It details the inclusion of relevant source and test files and generation of a project for use in the following exercise.

Exercise 3B — This exercise involves design optimization of a matrix multiplication function through the use of various directives. It presents the Vivado HLS design environment and method of synthesis and analysis of project solutions.

Exercise 3C — Finally, a more detailed look at how Vivado HLS synthesises interfaces is investigated.

Exercise 3A Creating Projects in Vivado HLS

In this exercise we will present the creation of Vivado HLS projects using both the Vivado HLS GUI and the use of Tcl scripting to expedite the process.

(a) Before we begin it is necessary to copy the files from **C:\Zynq_Book\sources\hls** to a new directory, **C:\Zynq_Book\HLS.**

(b) Launch the Vivado HLS GUI by navigating to **Start** > **All Programs** > **Xilinx Design Tools** > **Vivado 2015.1** > **Vivado HLS** > **Vivado HLS 2015.1**

(c) When the Vivado HLS GUI loads, you will be presented with the *Welcome* screen as in Figure 3.1.

Figure 3.1: Vivado HLS welcome screen

(d) Select the option to ***Create New Project*** in Figure 3.1

(e) At the Project Name dialogue, enter ***matrix_mult_prj*** as the ***Project name*** and ***C:\Zynq_Book\HLS\tut3A*** as ***Project location***.

Click ***Next***.

(f) You will now be prompted to add or remove source files for the project. All C-based source files for this tutorial have been created in advance, as we seek to guide through the design flow rather than the programming itself. Click ***Add Files...*** and navigate to **C:\Zynq_Book\HLS\tut3A**

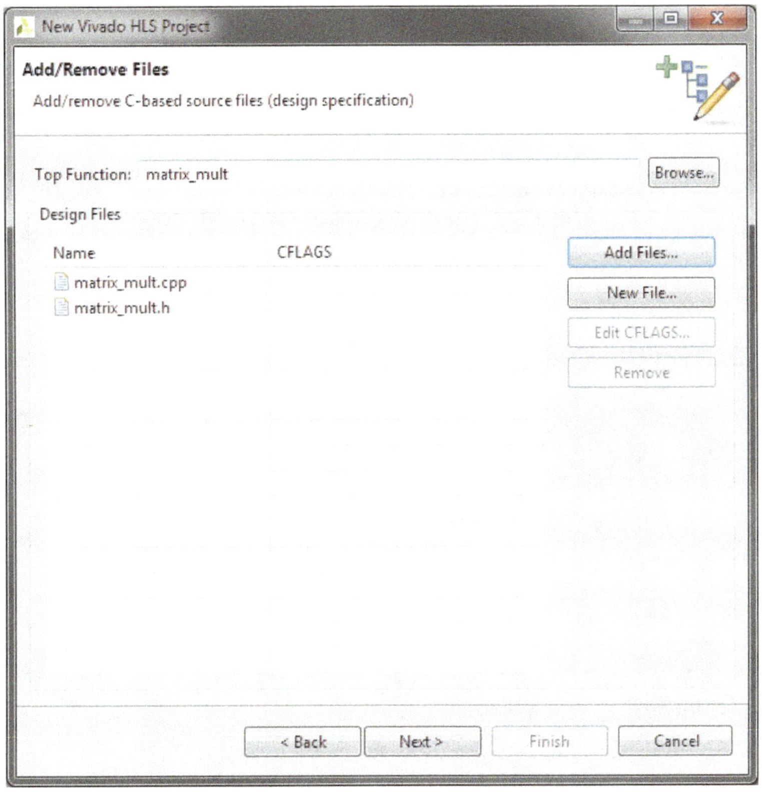

Figure 3.2: Adding files to a Vivado HLS project

Select the files ***matrix_mult.cpp*** and ***matrix_mult.h*** (hold down control to select multiple files) and click ***Open.*** Set the top function to ***matrix_mult*** as in Figure 3.2.

Click ***Next***.

(g) You will now be prompted to add a testbench file for design testing. Once more, click ***Add Files...*** and navigate to the previous directory this time adding the file ***matrix_mult_test.cpp*** and clicking ***Next***.

The next step is configuring a solution for a specific FPGA technology. In this case, leave the solution name and ensure the ***clock period*** is set to 5 as shown in Figure 3.3.

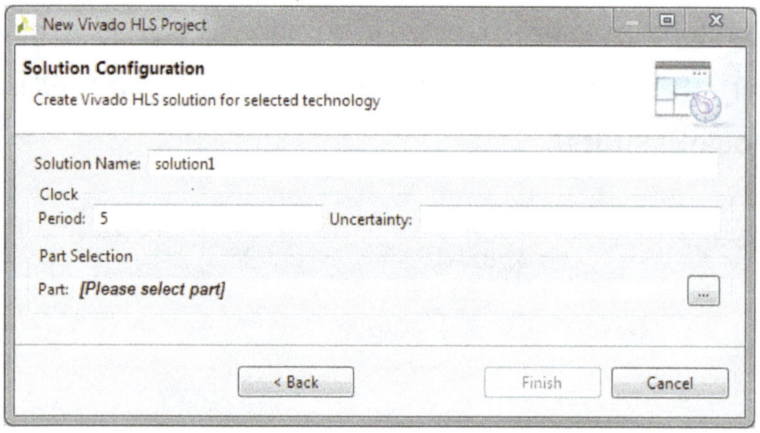

Figure 3.3: Solution Configuration Window

Zed

Since we are using the **ZedBoard** with the **Zynq-7020** chip click, ... in the part selection panel.

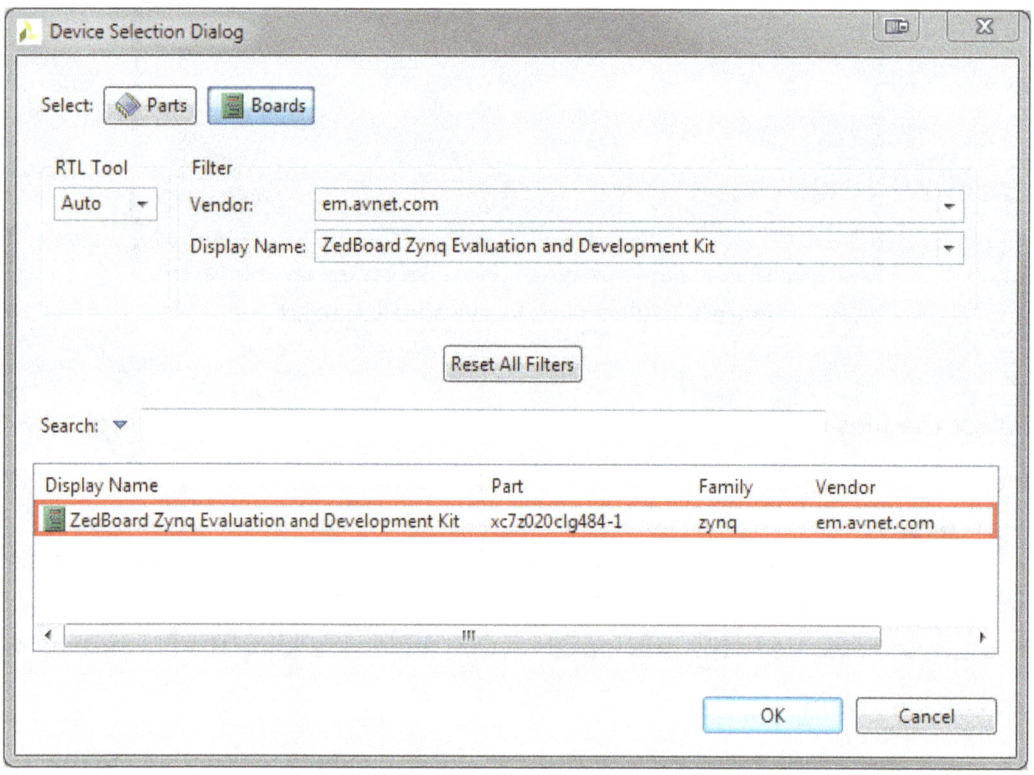

Figure 3.4: Zedboard device selection dialogue

In the ***Select*** section, click *Boards* and then filter the board parts using the filter drop down menus as in Figure 3.4. Select *ZedBoard Zynq Evaluation and Development Kit* and click ***OK.*** Click ***Finish***.

Zybo Since we are using the **Zybo** with the **Zynq-7010** chip click, [...] in the part selection panel.

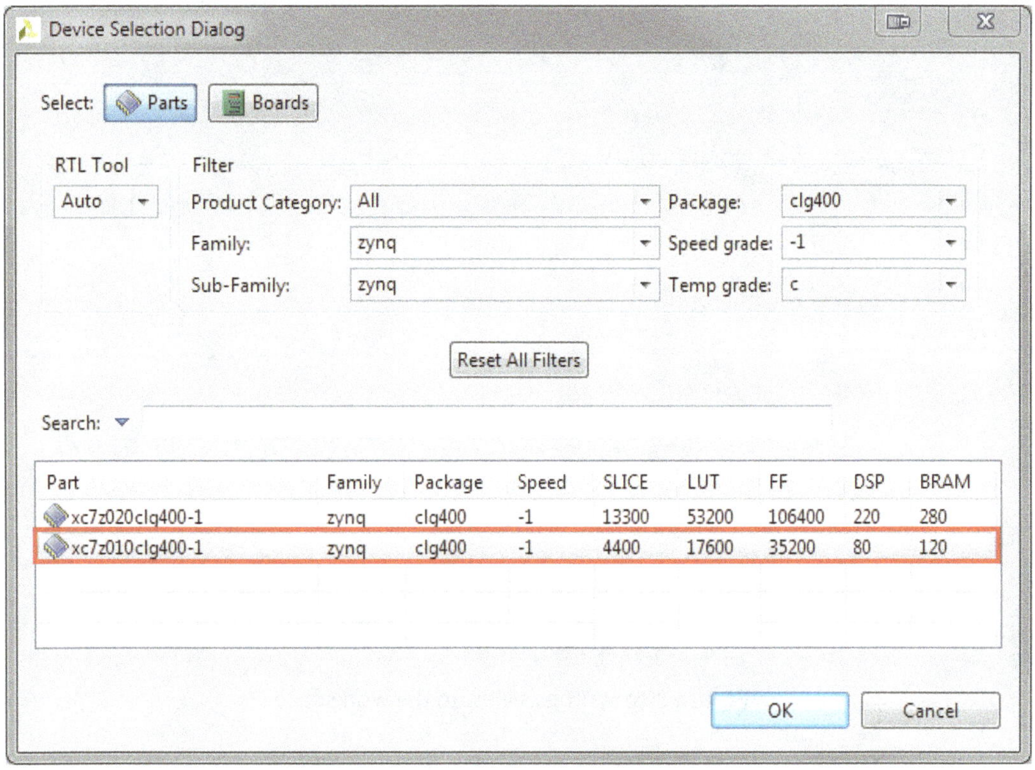

Figure 3.5: Zybo Part selection dialogue

In the ***Select*** section click *Parts* and then filter the board parts using the filter drop down menus, as shown in Figure 3.5. The required part can be confirmed by inspecting the Zynq chip on the Zybo development board. The **Z7010** Zynq chip with a **clg400** package should be selected. Click ***OK***.

Click ***Finish***.

Resume (h) The project will be generated and the workspace will open in ***Synthesis*** mode for the generated project and solution as in Figure 3.6.

Expanding the ***Source*** and ***Test Bench*** sections in the ***Explorer*** tab on the left side shows the inclusion of the source and test files from the previous steps. Double clicking on these files opens them in the editor view for examination and editing.

The project consists of a matrix multiplier, which multiplies two matrices *inA* and *inB* to produce the output *prod*. The testbench performs the multiplication of two known matrices and checks the value of *prod* against expected values.

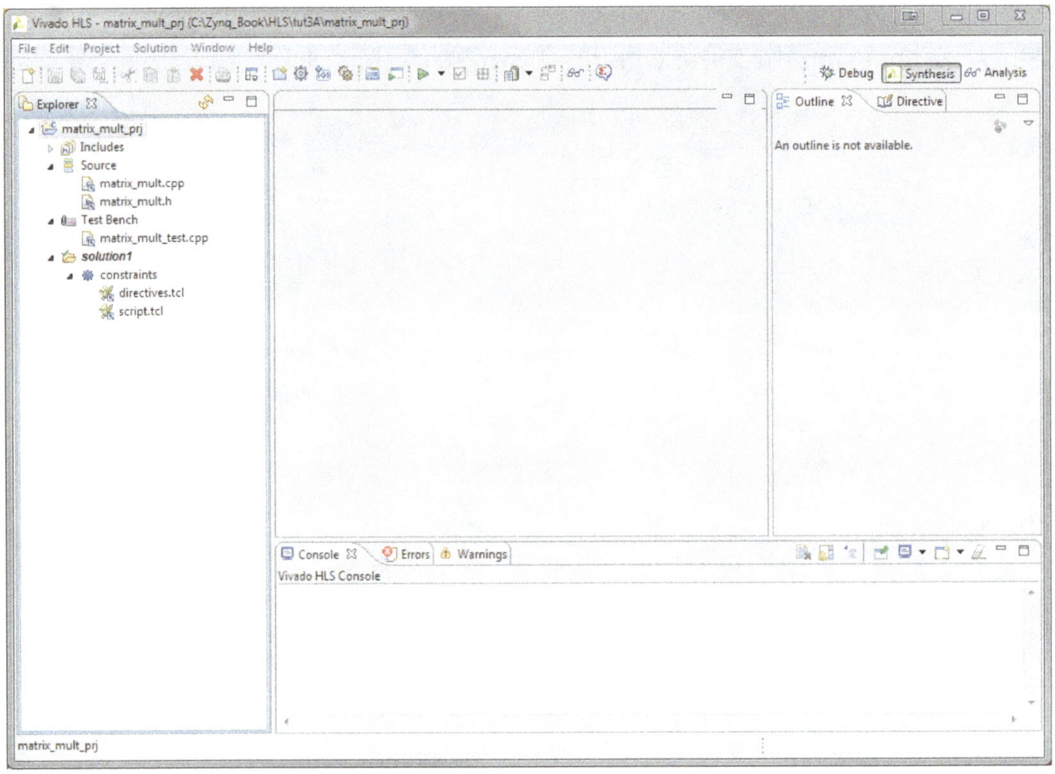

Figure 3.6: Synthesis view in the workspace

While the process of getting to this stage of HLS development is relatively straightforward, it can be quite repetitive and so can be facilitated by use of Tcl scripting. This automates the process of project naming and adding files. As such, we will now demonstrate the creation of the same project using the aforementioned scripting approach.

(i) First, close the Vivado HLS GUI. We will now open the Vivado HLS Command Prompt.

Launch the command prompt by navigating to **Start > All Programs > Xilinx Design Tools > Vivado 2015.1 > Vivado HLS > Vivado HLS 2015.1 Command Prompt**.

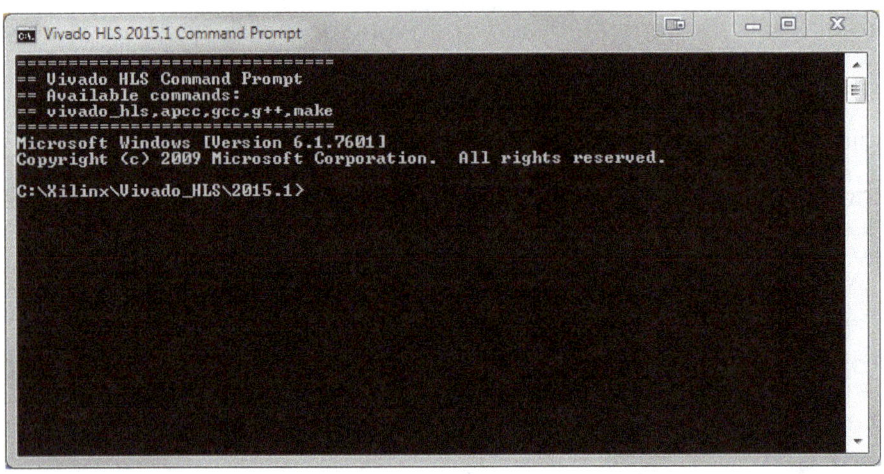

Figure 3.7: Vivado HLS command prompt

(j) It is observed that the default directory for commands is the install directory of Vivado HLS, as in Figure 3.7. To change this to the working directory for this tutorial, use the following commands, followed by pressing the ***Enter*** key.

- **cd..** — This is a change directory command which moves up a level in the directory. Repeat this until you have reached the level of the C: drive.
- **cd Zynq_Book** — This changes directory to the Zynq_Book folder.
- **cd HLS** — This changes directory to Zynq_Book/HLS.
- **cd tut3A** — This changes directory to Zynq_Book/HLS/tut3A.

The command prompt should now be in the working directory **C:\Zynq_Book\HLS\tut3A**. This folder contains the source and test files for a project, and also the Tcl script required to build the project, ***run_hls_zed.tcl*** and ***run_hls_zybo.tcl***

(k) With the correct working directory and the required files present in that directory, we can now build the project. This is achieved through simply running the Tcl script using the command:

Zed

```
vivado_hls -f run_hls_zed.tcl
```

Zybo

```
vivado_hls -f run_hls_zybo.tcl
```

Resume This will begin the process of creating the project and adding source and test bench files. A HLS solution is then created before configuring the project for the target device. Finally a C simulation is run which utilises the test bench to ensure the project operates correctly.

The testbench performs identical multiplications using the HLS hardware solution and software, and compares the results. If these results are identical, a *"Test passed!"* message is displayed:

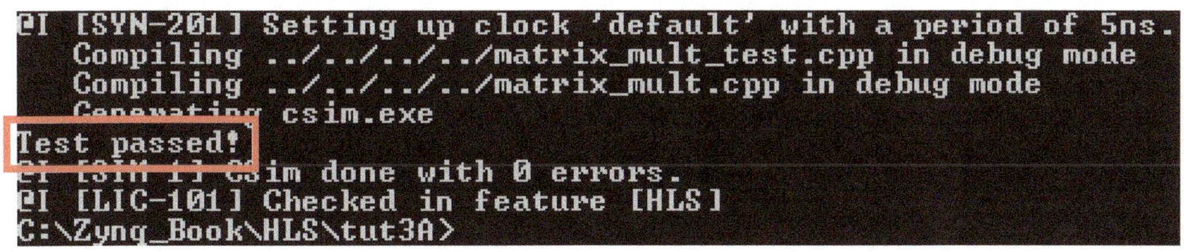

(l) To open the project in the Vivado HLS GUI enter the following command:

```
vivado_hls -p matrix_mult_prj
```

And press ***Enter***. This will open the Vivado HLS GUI for the project, which we will utilise in the next exercise.

Using the project generated in the previous exercise, we will now investigate the process of design optimisation in Vivado HLS. This will also provide an insight into the flow from project creation to C synthesis and C/RTL cosimulation. We will also discuss the use of the *Analysis* perspective in analysing a HLS solution.

Exercise 3B Design Optimisation in Vivado HLS

(a) You should already have the GUI open from the previous exercise, but if you don't, open the project ***matrix_mult_prj*** in the directory **C:\Zynq_Book\HLS\tut3A.**

(b) Expand the tabs for ***Source*** and ***Test Bench*** in the ***Explorer*** tab of the *Synthesis* view. As before, this shows that the source and test files have been successfully added to the project. Double clicking on each of these will open them in the editor allowing the code to be inspected and altered as required.

matrix_mult.cpp contains code that performs the multiplication of two matrices through use of iterative loops that run through the rows and columns of the matrices to calculate the product.

matrix_mult.h contains definitions and the prototype function for the matrix multiplication.

matrix_mult_test.cpp is the test bench file which calculates the product of two given matrices using both the HLS hardware solution and software, comparing to two to ensure successful operation.

(c) Click the ***Run C Simulation*** button in the toolbar to run a C simulation of the solution. Leave the options as default (no boxes checked, no input arguments) and click ***OK***. Upon completion of the simulation, the *"Test passed!"* message will be displayed in the console in the bottom of the screen as in Figure 3.8.

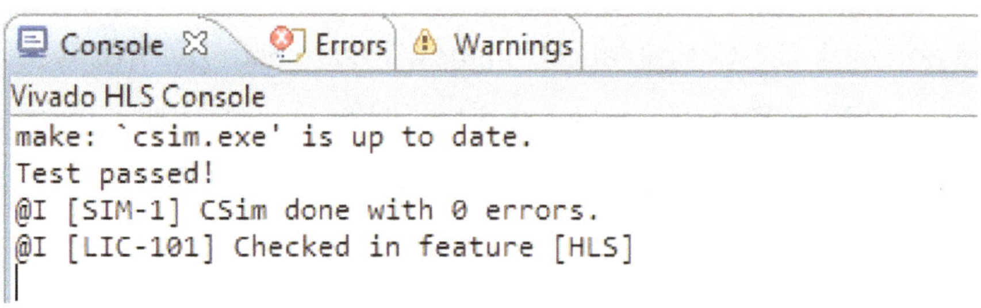

Figure 3.8: Vivado HLS console detailing successful testing

(d) The next step is to synthesise the C++ code using HLS. Click the ***C Synthesis*** button in the toolbar. Vivado HLS will begin the process of converting the C++ code into an RTL model with associated VHDL/Verilog/SystemC code. The console details the steps performed in achieving this.

Upon completion, a *Synthesis Report* will open automatically. This details various aspects of the synthesised design, such as information concerning timing and latency and FPGA resource utilisation estimates. (You may require to expand sub-sections to see results.)

Performance Estimates

⊟ **Timing (ns)**

⊟ **Summary**

Clock	Target	Estimated	Uncertainty
default	5.00	3.44	0.63

⊟ **Latency (clock cycles)**

⊟ **Summary**

Latency		Interval		
min	max	min	max	Type
686	686	687	687	none

⊟ **Detail**

⊞ **Instance**

⊟ **Loop**

	Latency			Initiation Interval			
Loop Name	min	max	Iteration Latency	achieved	target	Trip Count	Pipelined
- Row	685	685	137	-	-	5	no
+ Col	135	135	27	-	-	5	no
++ Product	25	25	5	-	-	5	no

The synthesised design has an interval of 687 clock cycles. Each input array contains 25 elements (as it uses 5x5 matrices) and so this suggests roughly 27 clock cycles per input read.

We can now run a C/RTL cosimulation to ensure that the synthesised RTL behaves exactly the same as the C++ code under test.

Click the ***Run C/RTL Cosimulation*** button . For the RTL selection, ensure **VHDL** is selected and click ***OK***. Cosimulation will now begin, with the RTL system being generated using VHDL. This process may take a short while to complete but progress can be viewed in the console. Upon completion, the *Cosimulation Report* will be opened as in Figure 3.10

Result

		Latency			Interval		
RTL	Status	min	avg	max	min	avg	max
VHDL	Pass	686	686	686	687	687	687
Verilog	NA	NA	NA	NA	NA	NA	NA

Export the report(.html) using the Export Wizard

Figure 3.10: Cosimulation report for the matrix multiplier, solution

Note the *"Pass"* message of Figure 3.10 indicating that the RTL behaves the same as the C++ source code.

(e) Create a new solution for the design by either clicking the ***New Solution*** button in the toolbar or the menu option ***Project*** > ***New Solution***. Click ***Finish*** to accept the defaults for *solution2*.

(f) Double click on *matrix_mult.cpp* in the ***Source*** section of the ***Explorer*** tab to ensure the code is visible in the workspace. We will now insert a directive which will pipeline the nested loops of the matrix multiplication code. This will perform loop flattening, removing the need for loop transitions.

Open the ***Directives*** tab to the right of the workspace. Click on ***Product*** and you will observe the associated portion of code highlighted in the editor, in this instance the multiplication of array elements to produce the product elements of the resulting matrix. Right click on ***Product*** and select *Insert Directive:*.

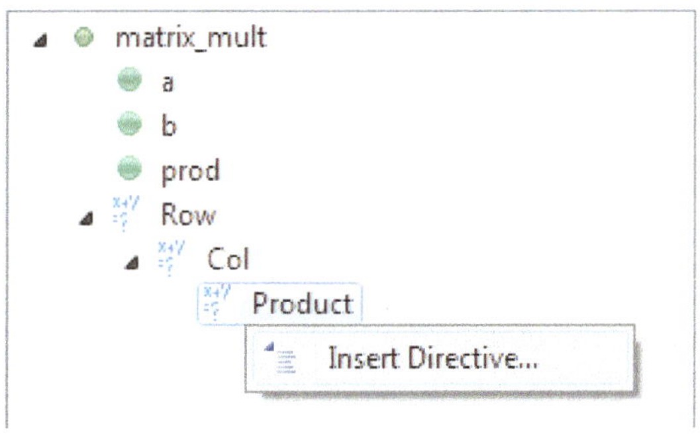

This will open the Directives Editor. Use the type drop-down menu to select the option *PIPELINE*. Click ***OK*** to accept the default options. The directives tab should now resemble Figure 3.11.

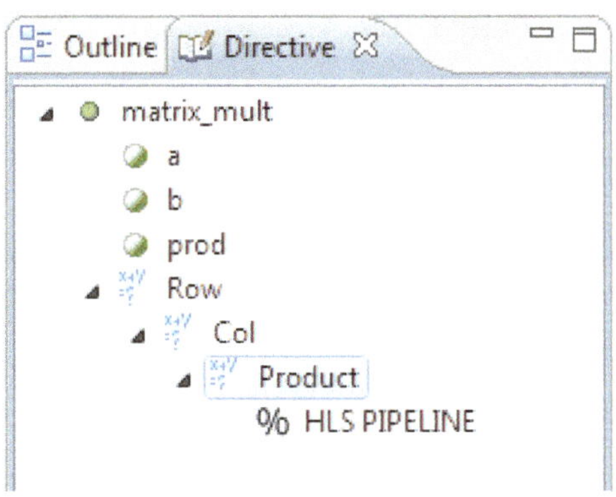

Figure 3.11: Pipelining nested loops in HLS

(g) Click the ***C Synthesis*** button to synthesise the RTL design. The console yields some information about the process of flattening the *Row* loop. It also explains that the default Initiation Interval (II) target of 1 could not be met for the *Product* loop. This is due to loop dependency.

Performance Estimates

⊟ **Timing (ns)**

⊟ **Summary**

Clock	Target	Estimated	Uncertainty
default	5.00	7.80	0.63

⊟ **Latency (clock cycles)**

⊟ **Summary**

Latency		Interval		
min	max	min	max	Type
426	426	427	427	none

⊟ **Detail**

⊞ **Instance**

⊟ **Loop**

	Latency			Initiation Interval			
Loop Name	min	max	Iteration Latency	achieved	target	Trip Count	Pipelined
- Row_Col	425	425	17	-	-	25	no
+ Product	12	12	5	2	1	5	yes

Figure 3.12: Synthesis report for the matrix multiplier, solution2

From the synthesis report shown in Figure 3.12 it is observed that the top level loop, *Row_Col* has not been pipelined as loop *Col* was not flattened. It is also observed an II of 2 was achieved despite the target of 1.

(h) Open the *Analysis* perspective by clicking on Analysis or ***Window > Analysis Perspective***. This will also open the *Performance* view showing how the various operations within the code are scheduled as clock cycles.

(i) Expand the loops *Row_Col* and *Product* by clicking on them to obtain the view shown in Figure 3.13.

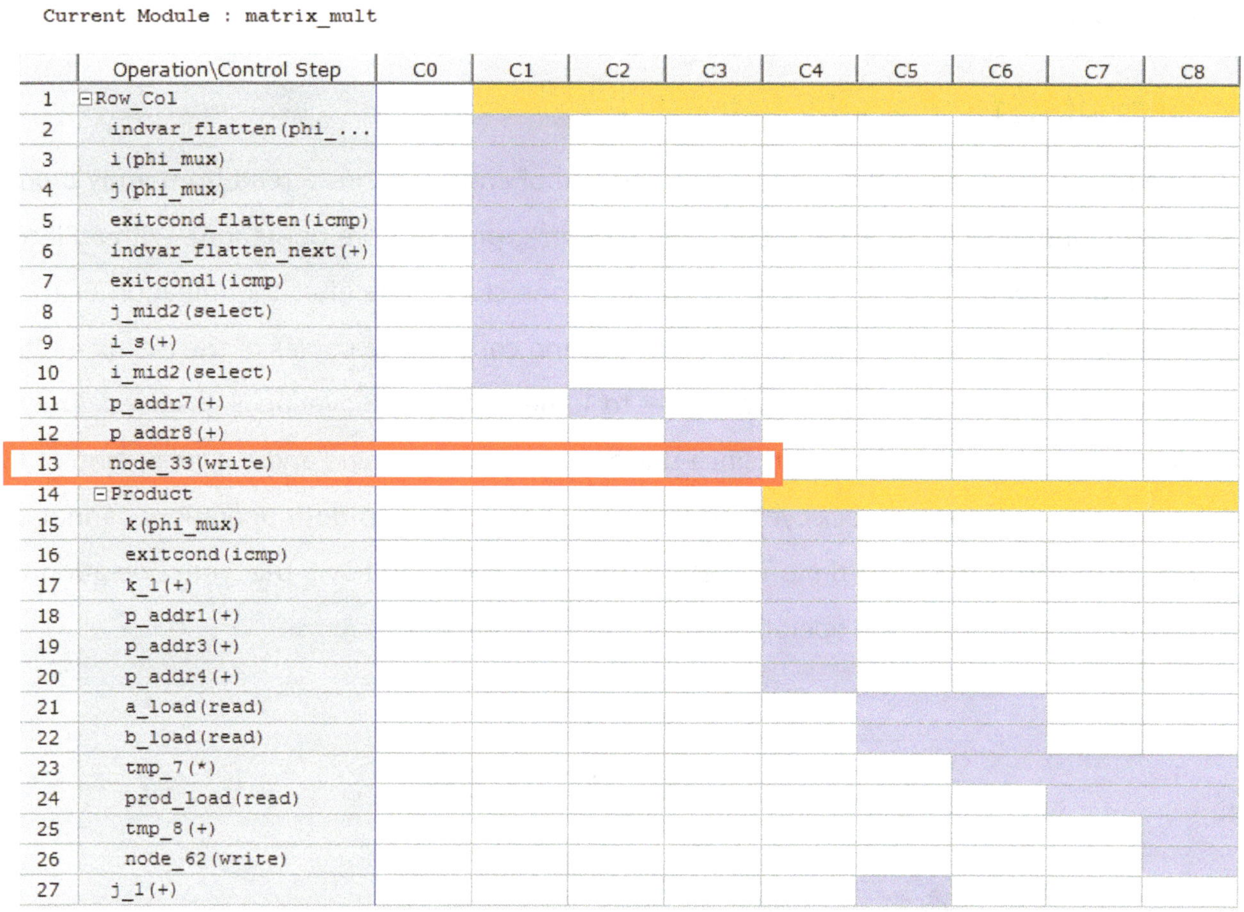

Figure 3.13: Performance view for solution2

Note that the highlighted write operation occurs in state C3, *node_33(write)*. Right clicking on this cell and selecting *Goto Source* will highlight the associated line of code in the source file. This is a write operation initialised as a write to a port in the RTL which occurs before any operations in the loop, *Product*, can be executed. This prevents the flattening of loop *Product* in to *Row_Col*.

Furthermore, the inability to meet the target of Initiation Interval (II) = 1 can be explained by considering consecutive iterations of the loop.

To show the console go to ***Window*** > ***Show View*** > ***Console.*** The console reveals the following message (you may need to scroll to find this message):

```
@W [SCHED-68] Unable to enforce a carried dependency constraint (II =
1, distance = 1) between 'store' operation (matrix_mult.cpp:16) of
variable 'tmp_8' on array 'prod' and 'load' operation ('prod_load',
matrix_mult.cpp16) on array 'prod'.
```

There exists a dependency between iterations of the operation at line 16 of the source code, which is the operation within the *Product* loop.

```
prod[i][j] += a[i][k] * b[k][j];
```

Due to the presence of the += operator, this line of code contains a read from array *prod* (the aforementioned load operation) and a write to array *prod* (a store operation). With an II of 1, a succeeding *Product* loop iteration would occur one clock cycle after the initiation of the first iteration. This is visualised in Figure 3.14 by pasting consecutive copies of the matrix multiplier operations, one above the other. With II set to 1, the highlighted overlap is observed. Arrays are mapped to BRAM by default, and since this overlap requires a *read* and a *write* operation to be performed on the same clock cycle, this is simply not possible as both operations cannot occur on the BRAM at the same time. Therefore, setting the II to 2 allows the *write* operation to be completed before the *read* operation of the next loop iteration begins.

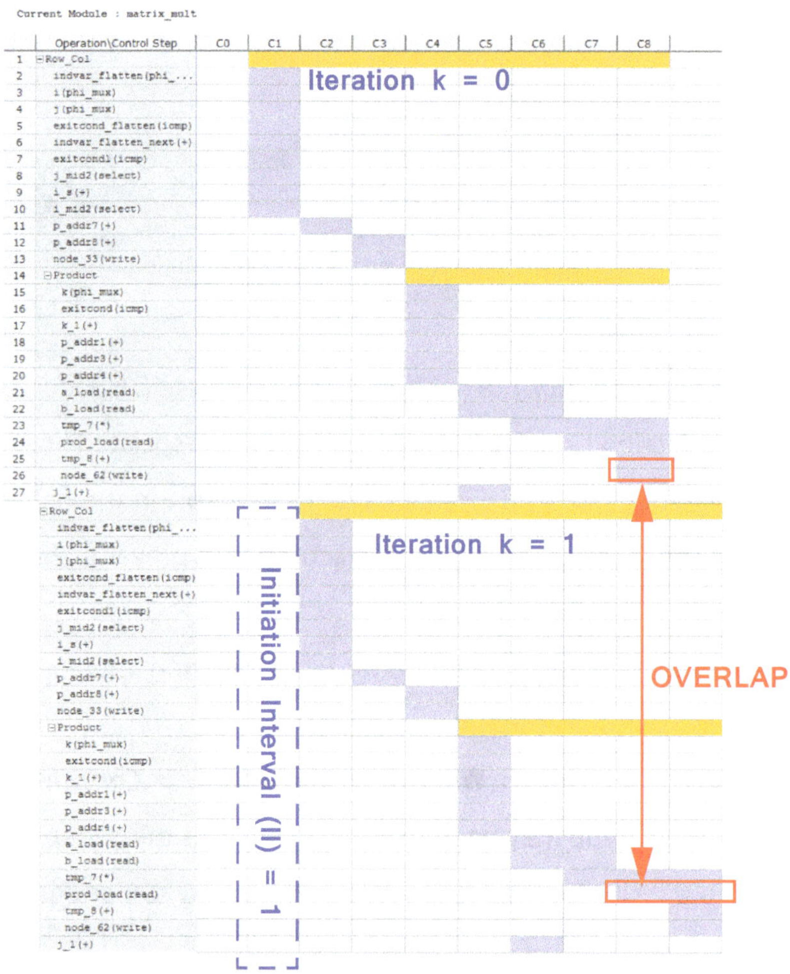

Figure 3.14: Consecutive iterations of Product loop with II = 1

(j) Return to the *Default Synthesis* perspective by selecting **Window > Synthesis Perspective**. Click **Yes** if a dialogue window appears.

We will now create a new solution which pipelines the *Col* loop, unrolling the *Product* loop to eliminate inter-iteration dependency but at the cost of increased operators and hence hardware cost.

(k) Create a new solution for the design by either clicking the ***New Solution*** button in the toolbar or the menu option ***Project*** > ***New Solution***. From the drop-down menus, ensure solution1 is selected, as this contains no existing directives or constraints.

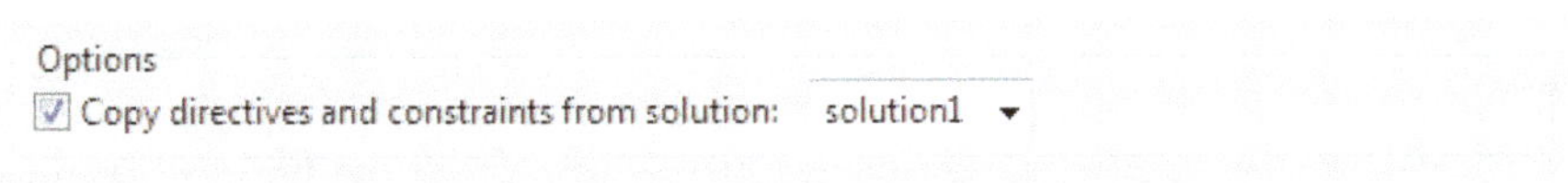

Click ***Finish*** to create the solution.

(l) Ensure the source code *matrix_mult.cpp* is visible in the editor. In the ***Directives*** tab, right-click on loop ***Col*** and select *Insert Directive*. From the drop-down menu, select directive type *PIPELINE*, ensure (II = 1) and click ***OK***. The directives tab should now resemble Figure 3.15.

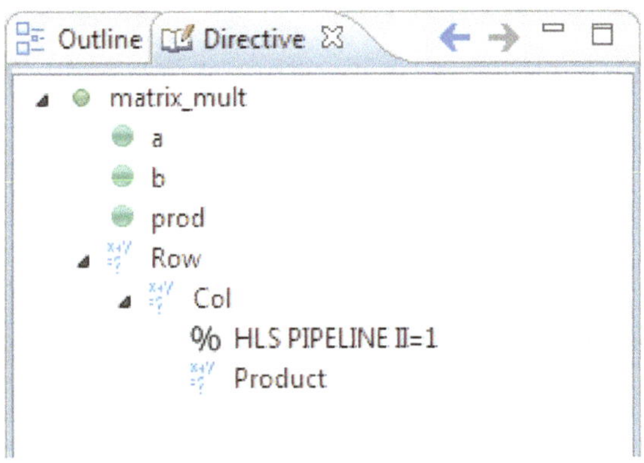

Figure 3.15: Pipelining Column Loop in HLS

(m) Click the ***C Synthesis*** button to synthesise the RTL design. Observing the *Console* will show that while *Product* was unrolled and loop *Row* was flattened. The II target of 1 could not be met for loop *Row_Col*, this time due to limitations in the resources. (You may need to scroll to locate this message).

```
@W [SCHED-69] Unable to schedule 'load' operation ('b_load_4',
matrix_mult.cpp:16) on array 'b' due to limited memory ports.
```

(n) Open the *Analysis* perspective by clicking on Analysis. This will open the *Performance* view. Switch to the *Resource* view by clicking the tab at the bottom of the screen.

(o) Expand the *Memory Ports* to view resource sharing on the memory within the system. Your view should look similar to Figure 3.16.

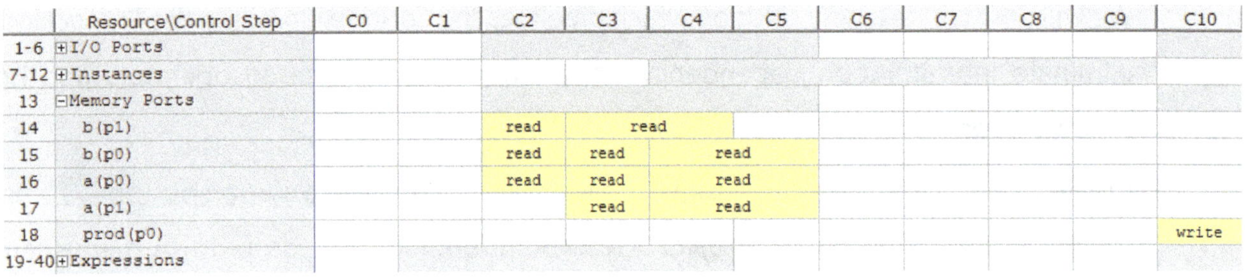

Figure 3.16: Resource sharing on memory ports of solution3

Figure 3.16 shows the operations per resource on each clock cycle. In actual fact, the 2 cycle read operation on *b* beginning in C3 overlaps with those in C4 so only a single cycle is visible. There are instances of both *a* and *b* being subjected to 3 read operations at once, which you will remember is not possible for dual-port BRAM. It is therefore necessary to partition these arrays into smaller sections, allowing modification of the array without altering the source code.

(p) Return to the *Synthesis* perspective by clicking on Synthesis .
Create a new solution for the design by either clicking the ***New Solution*** button in the toolbar or the menu option ***Project > New Solution***. Click ***Finish*** to accept the defaults for solution4. Note the directives will be copied from solution3.

For this solution, we will reshape the input arrays using directives. The ***Product*** loop is accessed via loop index *k*, therefore arrays ***a*** and ***b*** should be partitioned along their *k* dimension. Inspecting line 16 of *matrix_mult.cpp* it is observed that for ***a[i][k]*** this is dimension 2 and for ***b[k][j]*** dimension 1.

(q) Ensure the source code *matrix_mult.cpp* is visible in the editor, and open the ***Directives*** tab. Right-click on variable ***a*** and select *Insert Directive*. Ensure the directive is configured as in Figure 3.17, with *ARRAY_RESHAPE* selected as directive type and dimension specified as 2.

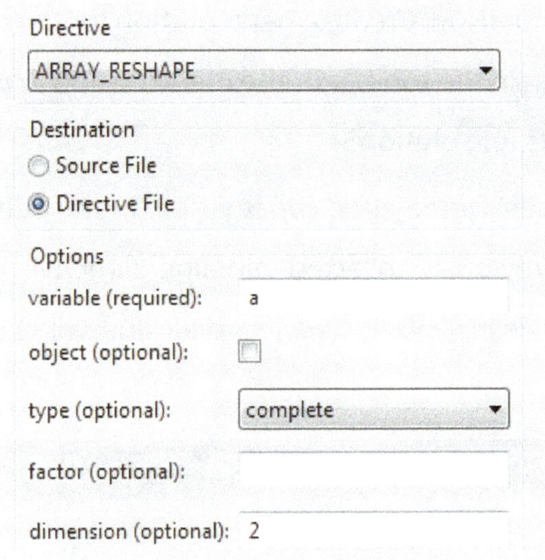

Figure 3.17: Directive configurations for reshaping array a

(r) Repeat for array ***b***, this time ensuring dimension is set to 1.

(s) Click the ***C Synthesis*** button to synthesise the RTL design. The synthesis report will open, showing that the target II of 1 has now been met.

⊟ **Latency (clock cycles)**

⊟ **Summary**

Latency		Interval		
min	max	min	max	Type
34	34	35	35	none

⊟ **Detail**

⊞ **Instance**

⊟ **Loop**

	Latency			Initiation Interval			
Loop Name	min	max	Iteration Latency	achieved	target	Trip Count	Pipelined
- Row_Col	32	32	9	1	1	25	yes

Figure 3.18: Synthesis report for solution4

The top-level of the design takes 34 clock cycles for completion, with the *Row_Col* loop outputting a sample after an iteration latency of 9. A sample is then read in every cycle (due to an II of 1), and after 25 counts all samples have been read in. The 34 clock cycles of this design is therefore justified by the 25 counts plus the latency of 9, as 25 + 9 = 34.

The function then proceeds to calculate the next set of data.

(t) The final optimisation in this exercise is to pipeline the function, rather than the loops within that function for comparison. Create a new solution for the design by either clicking the ***New Solution*** button in the toolbar or the menu option ***Project > New Solution***. Click ***Finish*** to accept the defaults for solution5.

(u) Ensure the source code *matrix_mult.cpp* is visible in the editor, and open the ***Directives*** tab. First, remove the previously inserted pipeline directive on loop *Col*. Right-click on the directive and select *Remove Directive*. If a dialogue window similar to that in Figure 3.19 appears, click ***No.***

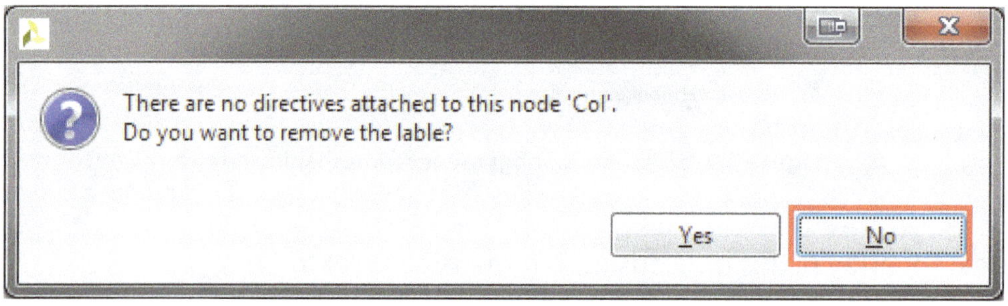

Figure 3.19: Node re-labelling dialogue window

(v) Right-click on the top level function *matrix_mult* and select *Insert Directive*. Select *PIPELINE* as the directive type and click ***OK***.

(w) Click the ***C Synthesis*** button to synthesise the RTL design.

Vivado HLS provides a tool for comparing synthesis reports. Click the button or the menu option ***Project > Compare Reports***.

Ensure solution4 and solution5 are added as in Figure 3.20. Click ***OK***.

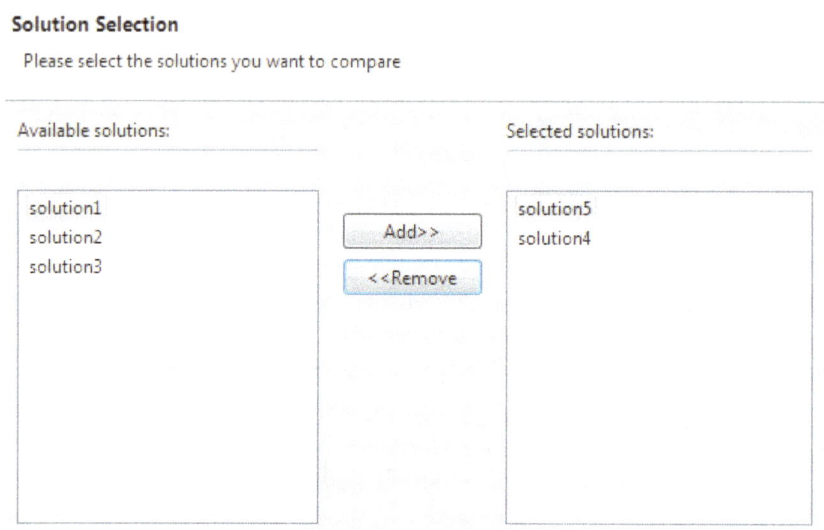

Figure 3.20: Solution selection for comparison

Figure 3.21 shows the comparison of synthesis report for solution4 (with loop pipelining) and solution5 (with top level function pipelining). It is observed that pipelining the top level function results in a design which reaches completion in fewer clocks, requiring only 13 clock cycles to begin a new transaction, rather than 35 for pipelining the loop.

However, this comes at the cost of increased hardware utilisation due to unrolling of all loops within the design. A trade-off is therefore necessary between system performance and the hardware utilisation of the design, and it is possible that a partially unrolled design may meet the performance requirements at a reduced hardware cost.

Performance Estimates

⊟ **Timing (ns)**

Clock		solution5	solution4
default	Target	5.00	5.00
	Estimated	4.35	4.35

⊟ **Latency (clock cycles)**

		solution5	solution4
Latency	min	24	34
	max	24	34
Interval	min	13	35
	max	13	35

Utilization Estimates

	solution5	solution4
BRAM_18K	0	0
DSP48E	125	5
FF	3511	244
LUT	546	82

Figure 3.21: Comparison of solution4 and solution5

The Zedboard, containing the **Z-7020** Zynq chip, accommodates 220 **DSP48E** slices which makes this device very suitable for implementing this hardware design. However the Zybo, which contains the **Z-7010** Zynq chip, only consists of 80 **DSP48E** slices. *Solution5* requires 125 **DSP48E** slices indicating that pipelining the top level function of the hardware design has increased device utilisation to the extent that the Zybo Zynq chip can no longer achieve full implementation. It is apparent that *solution4* would be suitable with reduced resource utilisation and increased latency.

(x) This completes the exercise. Close the Vivado HLS GUI.

We will now briefly explore the concept of interface synthesis in Vivado HLS, using the matrix multiplier function of the previous two exercises.

Exercise 3C Interface Synthesis

(a) Launch the command prompt by navigating to **Start > All Programs > Xilinx Design Tools > Vivado 2015.1> Vivado HLS > Vivado HLS 2015.1 Command Prompt**.

(b) Change the working directory to **C:\Zynq_Book\HLS\tut3C**. This folder contains the source and test files for a project, and also the Tcl script required to build the project, ***run_hls.tcl***.

(c) Run the Tcl script using the command:

Zed

```
vivado_hls -f run_hls_zed.tcl
```

Zybo

```
vivado_hls -f run_hls_zybo.tcl
```

Resume (d) To open the project in the Vivado HLS GUI enter the following command:

```
vivado_hls -p matrix_mult_prj
```

And press ***Enter***. This will open the Vivado HLS GUI for the project, which we will utilise in the next exercise.

(e) Open the source file *matrix_mult.cpp* from the ***Source*** section of the ***Explorer*** tab and click the ***C Synthesis*** button to synthesise the RTL design. When the synthesis report opens, scroll to the *Interface* section.

Note that the input arrays ***a*** and ***b***, and the resultant product array ***prod*** have been implemented using the

Interface

⊟ Summary

RTL Ports	Dir	Bits	Protocol	Source Object	C Type
ap_clk	in	1	ap_ctrl_hs	matrix_mult	return value
ap_rst	in	1	ap_ctrl_hs	matrix_mult	return value
ap_start	in	1	ap_ctrl_hs	matrix_mult	return value
ap_done	out	1	ap_ctrl_hs	matrix_mult	return value
ap_idle	out	1	ap_ctrl_hs	matrix_mult	return value
ap_ready	out	1	ap_ctrl_hs	matrix_mult	return value
a_address0	out	5	ap_memory	a	array
a_ce0	out	1	ap_memory	a	array
a_q0	in	8	ap_memory	a	array
b_address0	out	5	ap_memory	b	array
b_ce0	out	1	ap_memory	b	array
b_q0	in	8	ap_memory	b	array
prod_address0	out	5	ap_memory	prod	array
prod_ce0	out	1	ap_memory	prod	array
prod_we0	out	1	ap_memory	prod	array
prod_d0	out	16	ap_memory	prod	array

Figure 3.22: Interface summary for solution1

ap_memory protocol. This is inferred from the C++ source code, as the array type corresponds with the structure of memory.

Input arrays ***a*** and ***b*** are both 8 -bit signals on ports *a_q0* and *b_q0*. The output array, ***prod*** is a 16-bit signal on port *prod_d0*. Each signal has a corresponding 5-bit address port, designated as *a_address0, b_address0* and *prod_address0*.

The protocol also requires clock enable signals (*a_ce0* and *b_ce0*), and a write enable (*prod_we0)*.

Since the design requires more than one clock cycle to complete and is therefore synchronous, a clock and reset port have been synthesised as *ap_clk* and *ap_rst,* and both are 1-bit signals.

A block level control protocol with handshaking, *ap_ctrl_hs*, has also been implemented (*ap_start, ap_done, ap_idle* and *ap_ready)*.

- The *ap_start* input is asserted, prompting block operation. This produces three output control signals indicating the stage of operation.
- *ap_ready* indicates that the block is ready for new inputs.
- *ap_idle* is an indication that data is currently processing data.
- *ap_done* indicates that output data has been processed and is available.

Recalling Exercise 3B, the arrays were partitioned to reduce each into several smaller sections with expanded ports, control signals and implementation resources. This increased the bandwidth. This directly influenced the interface synthesis through use of directives.

This concludes this introduction to the design flow of Vivado HLS. This tool will be used further in future exercises, and synthesised RTL will be implemented as part of a larger functional model.

The Zynq Book Tutorials 4

IP Creation

v1.4, June 2015

Revision History

Date	Version	Changes
22/10/2013	1.0	First release for Vivado Design Suite version 2013.3
28/01/2014	1.1	Updated for changes in Vivado Design Suite version 2013.4
06/05/2014	1.2	Updated for changes in Vivado Design Suite version 2014.1
09/04/2015	1.3	Updated for changes in Vivado Design Suite version 2014.4
01/05/2015	1.3.1	Updated to include Zybo development board for Vivado Design Suite version 2014.4
19/06/2015	1.4	Updated for changes in Vivado Design Suite Version 2015.1

Introduction

The exercises in this tutorial will guide you through the process of creating custom IP modules, that are compatible with Vivado IP Integrator, from a variety of different sources. All created IP will be compatible with the Xilinx supported AXI-Lite interface, and will be connected as slave devices when implemented in Vivado IP Integrator.

All IP creation methods that are covered here coincide with those covered in the book:

- HDL
- MathWorks HDL Coder
- Xilinx Vivado HLS

The tutorial is split into three exercises, and is organised as follows:

Exercise 4A - In this exercise, HDL will be used to create a controller which will allow the LEDs on the Zynq development board to be controlled by software running on the PS. The Create and Package IP Wizard will be used to create an AXI-Lite interface wrapper which the LED control process and interface will be added to. The IP packaging process will then be used to create an IP block which is compatible with IP Integrator.

Exercise 4B - HDL Coder, the MathWorks HDL generation tool, will be explored in this exercise. A Least Mean Squares (LMS) adaptive filter will be created and tested in the Simulink workspace. The LMS design will then be used to generate HDL code by invoking the HDL Coder Workflow Advisor, where the option to generate a Xilinx IP Core will be selected. The various stages of the workflow will verify the design to ensure that it is HDL Coder compliant and produce the HDL code in a format that is compatible with IP Integrator. **Note:** You will require MATLAB, Simulink and HDL Coder in order to complete Exercise 4B.

Exercise 4C - In this final exercise, Vivado HLS will be used to create an IP core for a Numerically Controlled Oscillator (NCO). An existing C-code algorithm will be simulated for testing, and run through the various stages of synthesis in order to create an IP Integrator compatible IP core.

Exercise 4A Creating IP in HDL

With Zynq devices comprising of both PS and PL parts, most IP that is created to run in PL should be able to communicate with software running on the PS. This requires that IP should be packaged with an interface that is compatible with the PS (in this case the AXI interface).

When creating IP in HDL, Vivado provides a set of AXI interface templates which can be created and customised via the *Create and Package IP Wizard*. The wizard, as the name suggests, facilitates two major functions: the creation of AXI4 IP peripherals; and the packaging of existing source files into an IP package which is compatible with the IP Integrator tool.

In this exercise we will actually be making use of both of these features to firstly create an AXI4-Lite IP template to which we will add functionality to allow the LEDs on the Zynq development board to be controlled via a software application running on the Zynq PS. Once the functionality has been added to the template, the source files will be packaged into an IP Integrator compatible IP block which will be included in a simple Zynq processor system.

We will start by creating a new Vivado project.

(a) Launch Vivado by double-clicking on the Vivado desktop icon: , or by navigating to **Start > All Programs > Xilinx Design Tools > Vivado 2015.1 > Vivado 2015.1**

(b) Select ***Create New Project*** from the *Getting Started* screen.

(c) The *New Project* dialogue will open. Click ***Next***.

(d) At the Project Name dialogue, enter ***led_controller*** as the ***Project name*** and ***C:/Zynq_Book*** as ***Project location***.

Make sure that you select the option to ***Create project subdirectory***. Ensure that all options match Figure 4.1.

Figure 4.1: Vivado Project Name specification - led_controller

Click ***Next***.

(e) Select ***RTL Project*** at the *Project Type* dialogue, and ensure that the option ***Do not specify sources at this time*** is not selected:

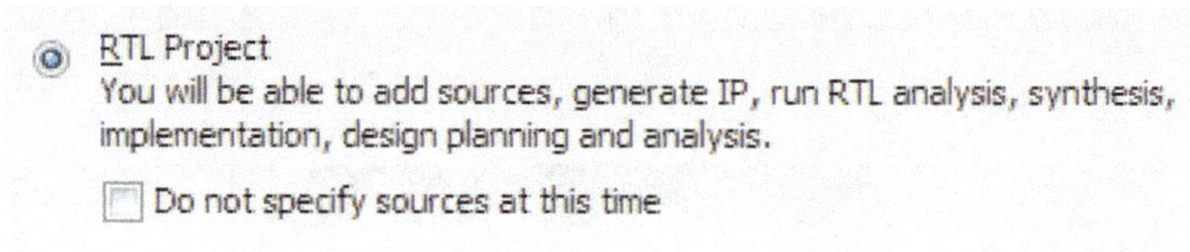

Click ***Next***.

(f) Select ***VHDL*** as the ***Target language*** in the *Add Sources* dialogue.

If existing sources, in the form of HDL or netlist files, were to be added to the project they could be imported at this stage.

As we do not have any sources to add to the project, click ***Next***.

(g) The *Add Existing IP (optional)* dialogue will open.

If existing IP sources were to be included in the project, they could be added here.

As we do not have any existing IP to add, click ***Next***.

(h) The *Add Constraints (optional)* dialogue will open.

This is the stage were any physical or timing constraints files could be added to the project.

As we do not have any constraints files to add, click ***Next***.

(i) The *Default Part* dialog will open. Here we will be selecting the Zynq part which we are targeting.

Zed

Select ***Boards*** from the *Specify* pane, ***ZedBoard Zynq Evaluation and Development Kit*** as the *Display Name*, and finally select the *Board Rev* which you have. In Figure 4.2 **version D** of the **ZedBoard** has been selected.

Click ***Next***.

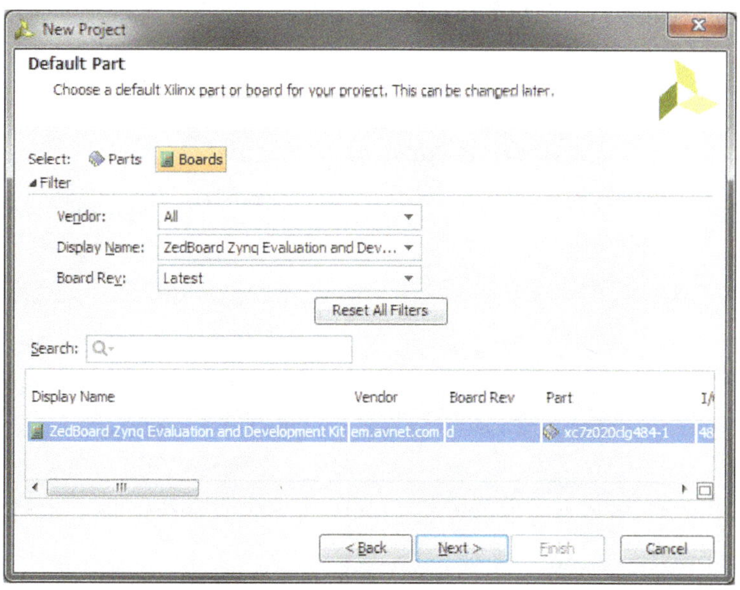

Figure 4.2: Zedboard Vivado Default Part dialogue

Zybo

Ensure you have carried out the Zybo board part set-up procedure at the beginning of Exercise 1A. Select ***Boards*** from the *Select* dialogue click ***Zybo*** from the *Display Name* list and **All** from the *Board Rev* list, as shown in Figure 4.3. Select the appropriate revision for your board (in this case **Rev. B.3** has been selected).

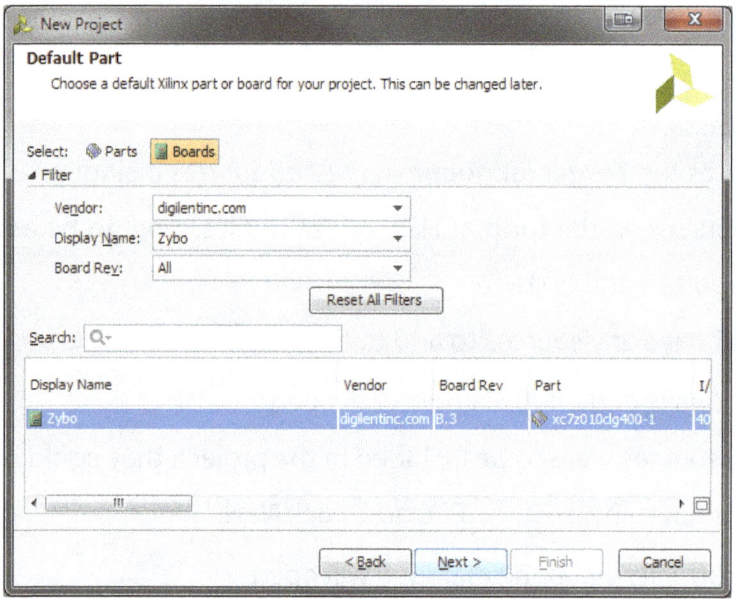

Figure 4.3: Zybo Default Part Dialogue Options

Click ***Next***.

Resume (j) Review the *New Project Summary* dialogue, and click ***Finish*** to create the project.

With the new project created, we can begin the process of creating our HDL-based IP.

(k) From the menu bar, select **Tools > Create and Package IP...**, as in Figure 4.4, to launch the ***Create and Package IP Wizard***.

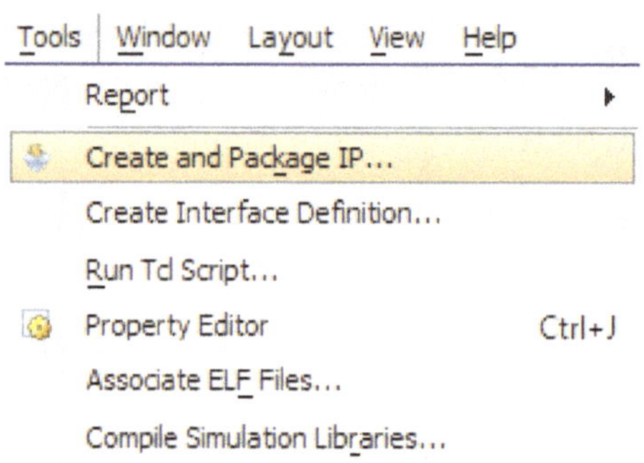

Figure 4.4: Create and Package IP menu bar selection

(l) The *Create and Package IP Wizard* dialogue will launch, as shown in Figure 4.5.

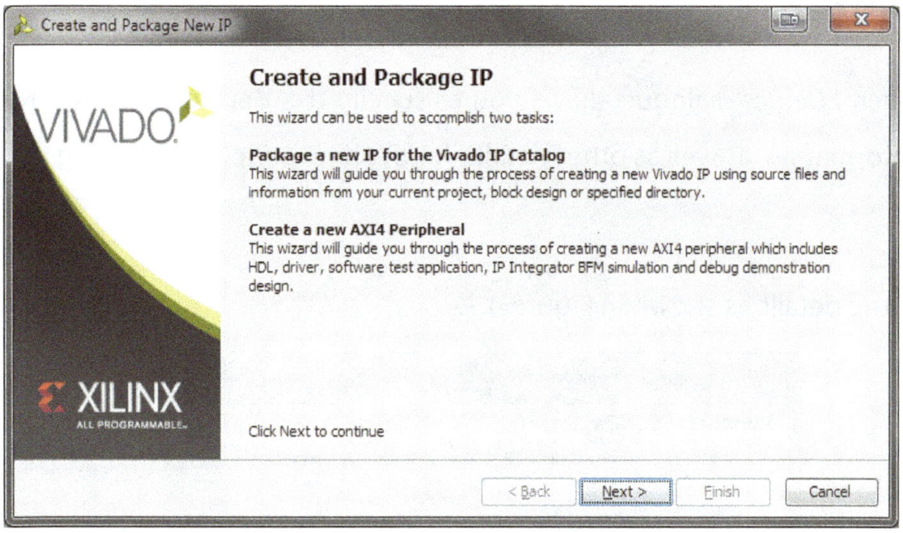

Figure 4.5: Create and Package IP Wizard dialogue

Click ***Next***.

The *Choose Create Peripheral or Package IP* dialogue (Figure 4.6) is where we specify whether to create a new peripheral template file or to package existing source files into an IP core.
In our case we want to create a new IP template.

(m) Select ***Create new AXI peripheral***, as shown in Figure 4.6.

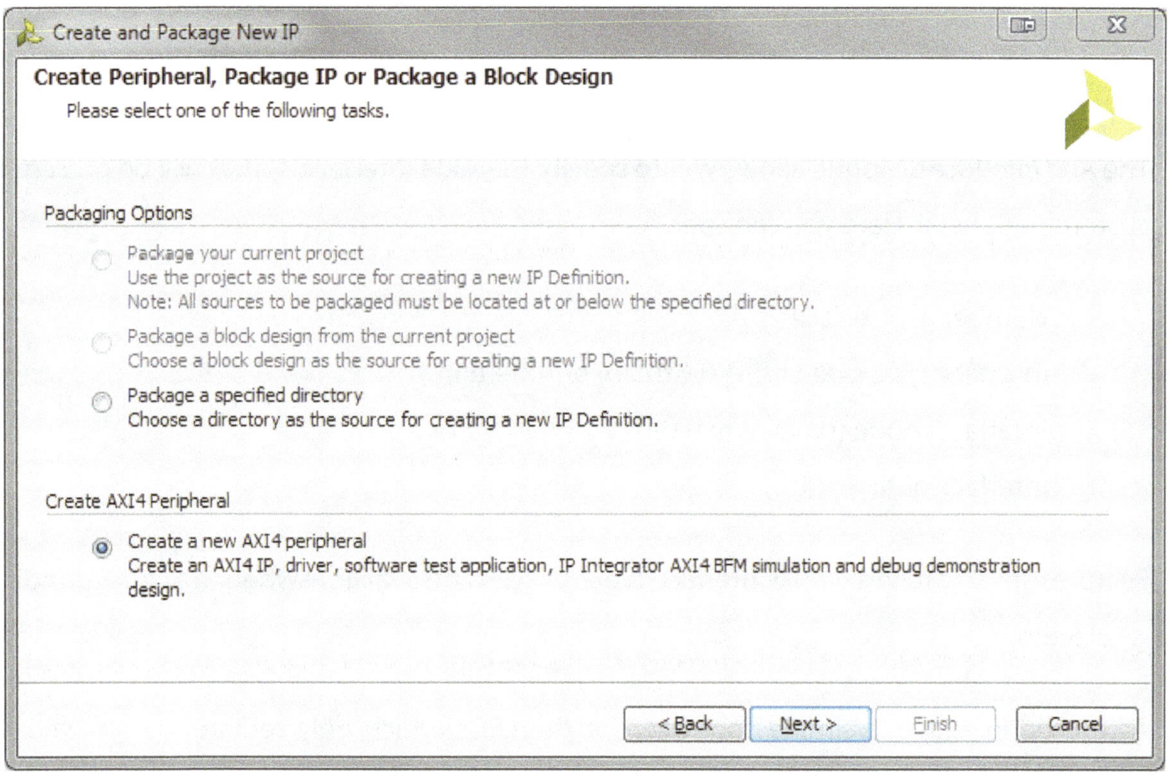

Figure 4.6: Choose Create or Package IP dialogue

Click ***Next***.

The *Peripheral Details* dialogue allows you to specify the **Vendor, Library, Name and Version (VLNV)** information, as well as other details, for the new peripheral, leaving the *IP Location* as the default.

(n) Fill in the details as shown in Figure 4.7.

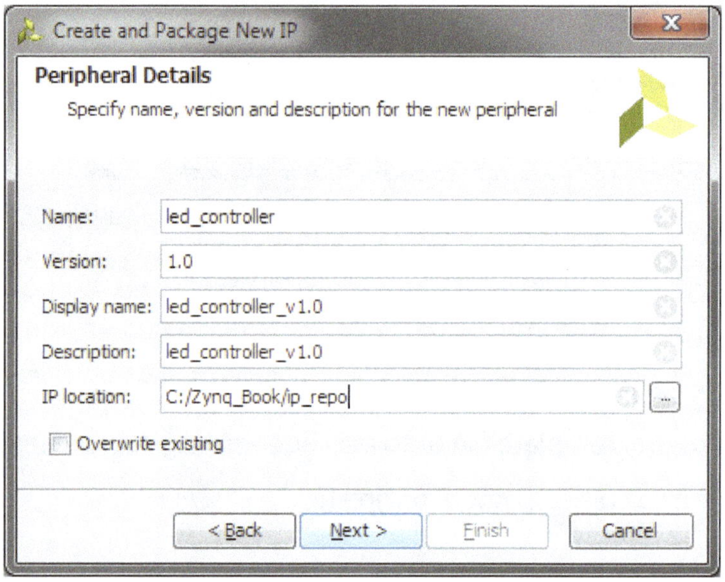

Figure 4.7: Peripheral Details dialogue

Click ***Next***.

The *Add Interface* dialogue allows you to specify the AXI4 interface(s) that will be present in your custom peripheral. Here you can specify:

- Number of interfaces
- Interface type (AXI-Lite, AXI-Stream or AXI-Full)
- Interface mode (slave or master)
- Interface data width

Features specific to individual interface types will also be available when the corresponding type is selected.

As our peripheral is a simple controller for the LEDs which only requires single values to be transferred to it, an AXI-Lite slave interface is sufficient. Only one memory mapped register is required for our simple controller, but as the minimum number that can be specified in the dialogue is 4, we will choose that.

(o) Specify the *Add Interface* dialogue as shown in Figure 4.8.

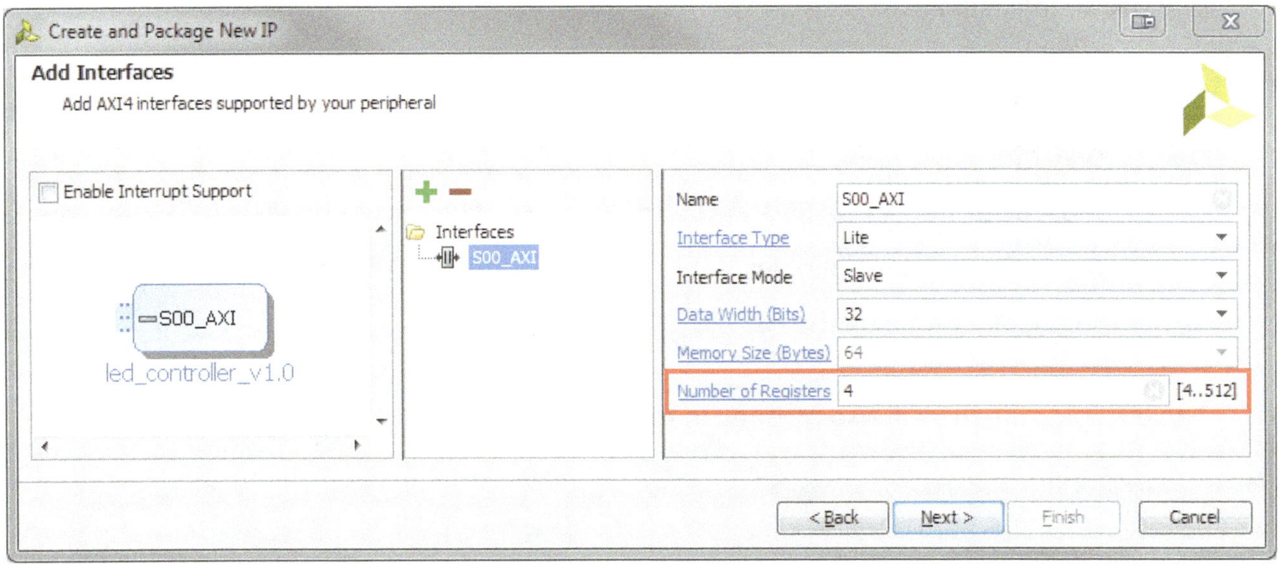

Figure 4.8: Add Interface dialogue

Click ***Next***.

(p) Review the information in the *Create Peripheral* dialogue, which details the output files which will be created.

Select the option to ***Edit IP***. This will create the IP peripheral files and create a new Vivado project where the functionality of the peripheral can be modified in the source HDL code, and then packaged.

Click ***Finish*** to close the *Wizard* and create the peripheral template.

A new Vivado project, named **edit_led_controller_v1_0**, will open.

In the *Sources* pane, you should see two HDL source files (you may need to expand the file selection):

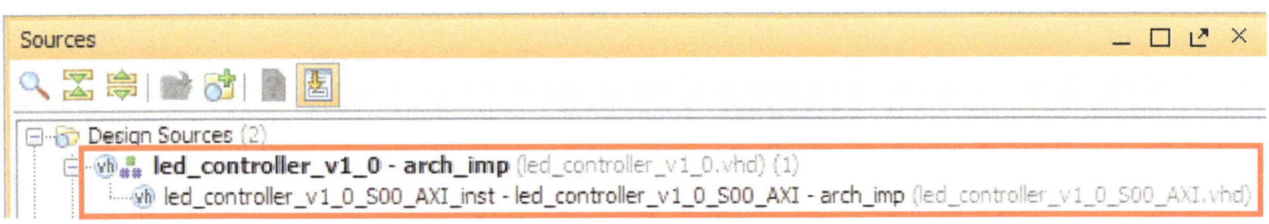

As we specified our target language as **VHDL** in Step (f) earlier, the template files have been generated in VHDL. Had we specified Verilog as the target language, Verilog source files would have been created.

The two source files are:

- **led_controller_v1_0.vhd** — This file instantiates all AXI-Lite interfaces. In this case, only one interface is present.
- **led_controller_v1_0_S00_AXI.vhd** — This file contains the AXI4-Lite interface functionality which handles the interactions between the peripheral in the PL and the software running on the PS.

The *IP Packager* pane will also be open in the *Workspace*:

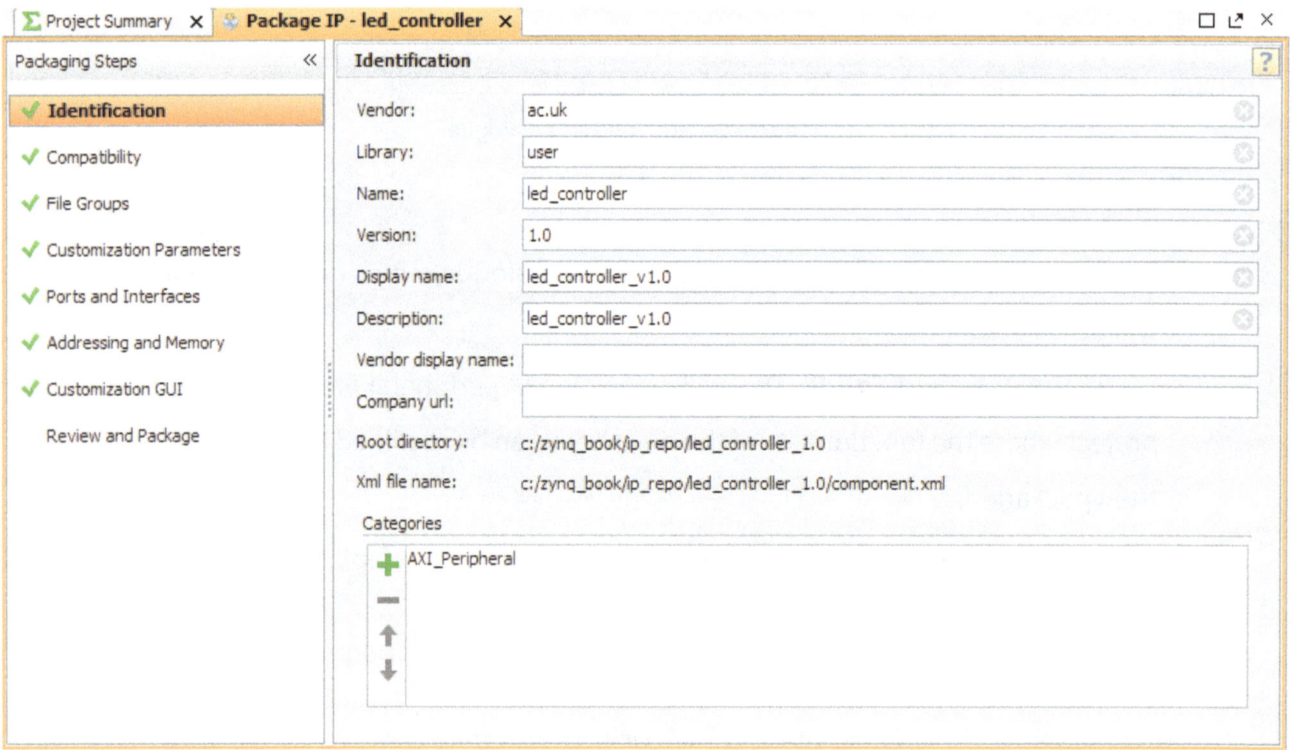

The information that we specified about our peripheral in Step (n) will be visible. The *Vendor* parameter will be dependant on your computers network domain and can be changed.

We can now add the functionality to our **led_controller** peripheral. We will be adding a new output port to the peripheral template to allow it to connect to the LED pins on the Zynq device, as well as assigning the value received from the Zynq PS to the new output port.

(q) Open ***led_controller_v1_0_S00_AXI.vhd*** by double-clicking on it in the *Sources* pane. The file will open in the *Workspace*.

Zed

Scroll down until you see the following comment in the entity port declaration:

```
-- Users to add ports here
```

Add the following port definition directly below the comment:

```
LEDs_out : out std_logic_vector(7 downto 0);
```

This creates a new output port with a width of 8-bits (a single bit to represent each of the LEDs on the ZedBoard).

Scroll to the bottom of the file. You should see the following comment:

```
-- Add user logic here
```

and add the following port/signal assignment:

```
LEDs_out <= slv_reg0(7 downto 0);
```

This assigns the value that is received from the Zynq PS (stored in the signal slv_reg0) to the output port that we created in the previous step.

Zybo

Scroll down until you see the following comment in the entity port declaration:

```
-- Users to add ports here
```

Add the following port definition directly below the comment:

```
LEDs_out : out std_logic_vector(3 downto 0);
```

This creates a new output port with a width of 4-bits (a single bit to represent each of the LEDs on the Zybo).

Scroll to the bottom of the file. You should see the following comment:

```
-- Add user logic here
```

and add the following port/signal assignment:

```
LEDs_out <= slv_reg0(3 downto 0);
```

This assigns the value that is received from the Zynq PS (stored in the signal slv_reg0) to the output port that we created in the previous step.

Resume (r) Save the file by selecting **File > Save File** from the Menu Bar, or using the keyboard shortcut **Ctrl+S**.

(s) Open ***led_controller_v1_0.vhd*** by double-clicking on it in the *Sources* pane. The file will open in the *Workspace*.

We must once again create a new output port to the top-level source file, and map it to the equivalent port that we created in the AXI4-Lite interface file in the previous steps.

Zed

Scroll down until you see the following comment in the entity port declaration:

```
-- Users to add ports here
```

and add the following port definition directly below the comment:

```
LEDs_out : out std_logic_vector(7 downto 0);
```

As we added a new port to the AXI4-Lite interface file, we must also add it to the component declaration in the top-level file.

Scroll down until you see the comment:

```
-- component declaration
```

A few lines further down you will see the component port declaration:

```
port (
```

Inside the port declaration (below the "`port (`" line), add the following output port definition:

```
LEDs_out : out std_logic_vector(7 downto 0);
```

Zybo

Scroll down until you see the following comment in the entity port declaration:

```
-- Users to add ports here
```

and add the following port definition directly below the comment:

```
LEDs_out : out std_logic_vector(3 downto 0);
```

As we added a new port to the AXI4-Lite interface file, we must also add it to the component declaration in the top-level file.

Scroll down until you see the comment:

```
-- component declaration
```

A few lines further down you will see the component port declaration:

```
port (
```

Inside the port declaration (below the "`port (`" line), add the following output port definition:

```
LEDs_out : out std_logic_vector(3 downto 0);
```

Resume Finally, we must add a port mapping between the LED output ports of the top-level file and the AXI4-Lite interface file.

(t) Scroll down until you see the comment:

```
-- Instantiation of Axi Bus Interface S00_AXI
```

A few lines further down you will see the component port map:

```
port map (
```

Inside the component port map (below "`port map (`" line), add the following port map:

```
LEDs_out => LEDs_out,
```

(u) Save the file.

Now that we have made the necessary modifications to the peripheral source files, we must repackage the IP to merge the changes.

(v) Return to **IP Packager** by selecting the ***Package IP - led_controller*** tab in the *Workspace*:

IP Packager will detect the changes to the source files, and the areas which need refreshed will be highlighted with the following icon: . You should see that the following two areas of interest need refreshed:

Customization Parameters

Ports and Interfaces

(w) Select ***Customization Parameters*** in the *IP Packager* pane.

You should see the following information message at the top of the pane:

Merge changes from Customization Parameters Wizard

Click ***Merge changes from Customization Parameters Wizard***

This will update the IP Packager information to reflect the changes made in the HDL source files.

NOTE: This process updates IP Packager information for all areas. You should see that the area of Ports and Interfaces no longer needs updated, and the icon has now been removed.

To verify that IP Packager has updated the Ports and Interfaces area, we will open it and check.

(x) Select ***Ports and Interfaces*** from the *IP Packager* pane.

Zed

You should notice that the LEDs_out port that we added to the source files has been added to the IP Ports pane and has a length of 8:

Name	Int...	Enablement Dependency	Is Declaration	Direction	Driver Value	Size Left	Size Left Dependency	Size Right	Size Right Dependency
S00_AXI	slave		☐						
Clock and Reset Signals			☐						
LEDs_out			☐	out		7		0	

Zybo

You should notice that the LEDs_out port that we added to the source files has been added to the IP Ports pane and has a length of 4:

Name	Int...	Enablement Dependency	Is Declaration	Direction	Driver Value	Size Left	Size Left Dependency	Size Right	Size Right Dependenc
S00_AXI	slave		☐						
Clock and Reset Signals			☐						
LEDs_out			☐	out		3		0	

Resume

The final step in creating our new IP peripheral, is to package the IP.

(y) Select ***Review and Package*** form the *IP Packager* pane.

(z) In the *After Packaging* panel, click ***edit packaging settings*** at the bottom:

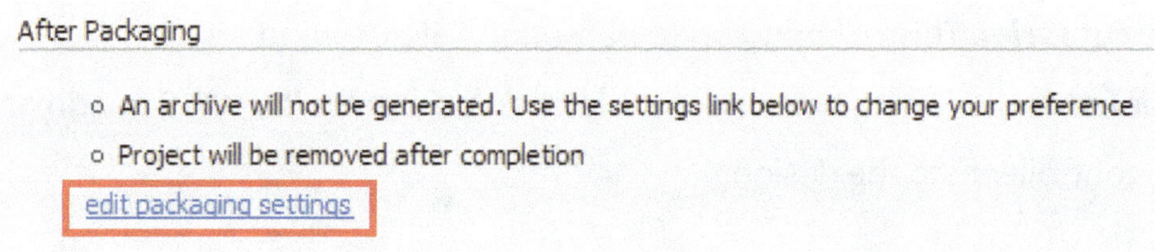

(aa) In the *Automatic Behaviour* panel, enable the option to ***Create archive of IP, Close IP Packager window*** and to ***Add IP to the IP Catalog of the Current Project.*** You may ***Delete project after packaging*** if you wish (does not have an impact on the remainder of this tutorial).

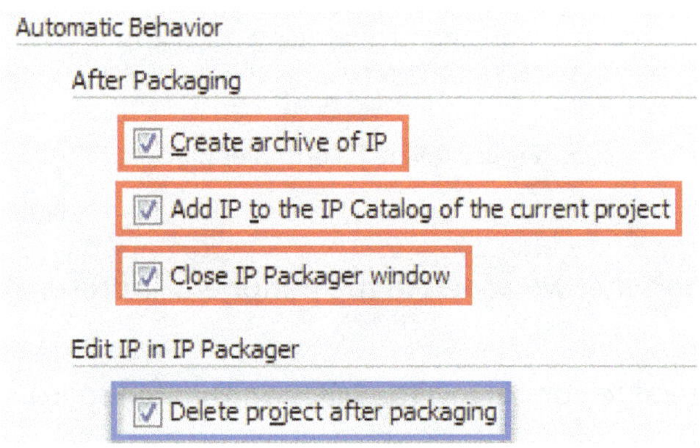

This makes a ZIP file archive of the packaged IP and close IP Packager once finished.

(ab) Click ***OK*** to apply the setting.

(ac) Review the information provided in the *Review and Package* window, and click ***Re-Package IP***.

(ad) A dialogue box will appear asking if you want to close the project, click **Yes.**

(ae) The changes made to the IP peripheral will be included in the repackaged IP, and the Vivado project will close.

We will now return to our original Vivado project, and create a simple Zynq processor block design to check that the functionality of our LED controller peripheral.

To start, we will create a new Block Design and add the IP peripheral which we just created to the design.

(af) In the *Flow Navigator* window, select ***Create Block Design*** from the *IP Integrator* section. Enter ***led_test_system*** in the *Design name* box, and click ***OK*** to create the blank design.

(ag) Right-click anywhere in the blank canvas, and select ***Add IP***. Alternatively, use the keyboard shortcut ***Ctrl+I***. This will bring up to pop-up IP Catalog window.

Enter ***led*** in the *Search* box, and double-click ***led_controller_v1.0*** to add an instance of the LED controller IP to the design.

Zed

An led_controller_v1_0 block will now be present in the block design, as shown in Figure 4.9.

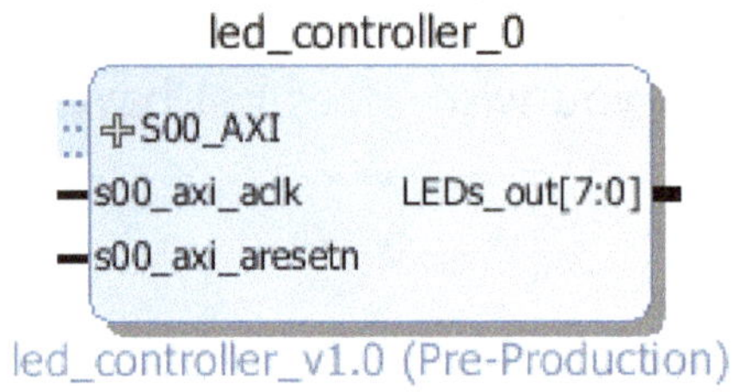

Figure 4.9: led_controller block

The 8-bit **LEDs_out** port that we added to the peripheral is present on the right side of the block.

To enable the peripheral to connect to the LEDs on the ZedBoard, we must make the **LEDs_out** port external. This allows the output port to be connected to specific physical pins on the Zynq device, which are connected to the LEDs.

Hover the mouse pointer over the **LEDs_out** interface (the little black stub next to the interface name) on the **led_controller** block until the cursor changes to a pencil. Right-click and select ***Make External***. Alternatively, select the interface and use the keyboard shortcut ***Ctrl+T***.

The block design should now resemble Figure 4.10.

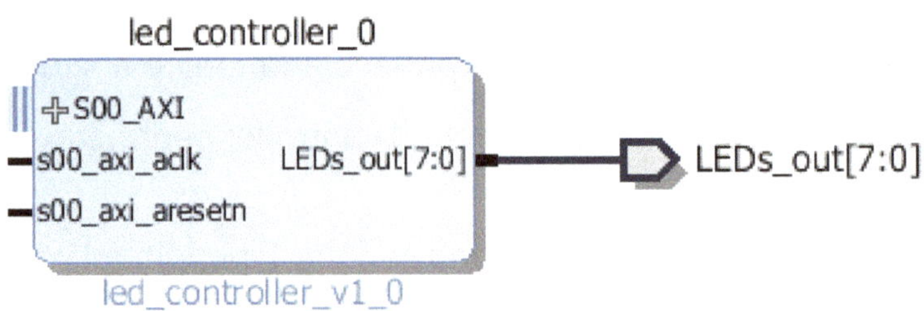

Figure 4.10: led_controller block with external port

Zybo

An led_controller_v1_0 block will now be present in the block design, as shown in Figure

4.11. The 4-bit **LEDs_out** port that we added to the peripheral is present on the right side of the block.

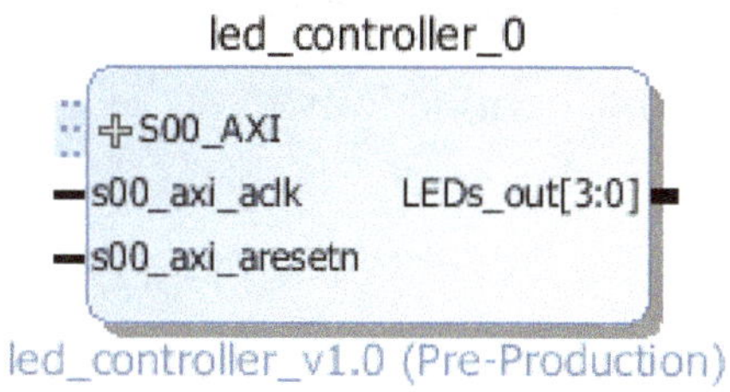

Figure 4.11: led_controller block

To enable the peripheral to connect to the LEDs on the Zybo, we must make the **LEDs_out** port external. This allows the output port to be connected to specific physical pins on the Zynq device, which are connected to the LEDs.

Hover the mouse pointer over the **LEDs_out** interface (the little black stub next to the interface name) on the **led_controller** block until the cursor changes to a pencil. Right-click and select ***Make External***. Alternatively, select the interface and use the keyboard shortcut ***Ctrl+T***.

The block design should now resemble Figure 4.12.

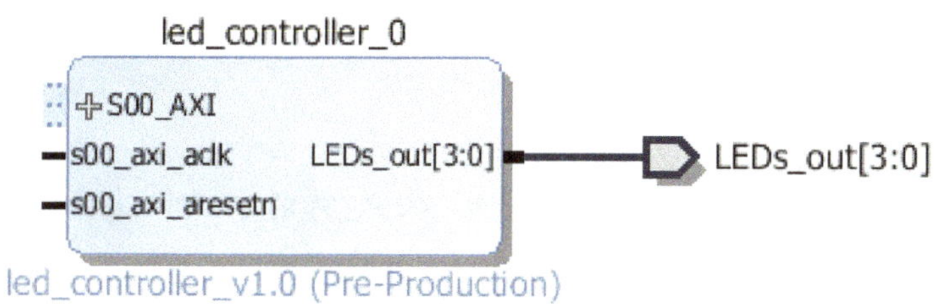

Figure 4.12: led_controller block with external port

Resume The next step is to add a Zynq Processing System block so that the LED Controller can be connected to it.

Add an instance of the **Zynq7 Processing System**, using the same procedure as in Step (ag). The *Designer Assistance* message at the top of the canvas will appear:

Click ***Run Block Automation***.

An information message will appear. Ensure that ***Apply Board Preset*** is selected, and click ***OK***.

This will make all necessary modifications to the Zynq processing system that relate to the board preset and make required external connections.

We must now connect the LED Controller to the Zynq Processing System. This step can also be carried out using Designer Assistance.

(ah) In the *Designer Assistance* message, click ***Run Connection Automation.***

An information message will appear, select ***led_controller_0/S00_AXI*** and click ***OK***.

This will add some additional blocks to the design which are required to connect the LED Controller to the Zynq Processing System.

Our block design is now complete.

(ai) Validate the design by selecting **Tools > Validate Design** from the *Menu Bar*. Alternatively, select the ***Validate Design*** button, , from the *Main Toolbar*, or use the keyboard shortcut ***F6***.

Dismiss the *Validate Design* message by clicking ***OK***.

We can now generate the HDL files for the design.

(aj) In the *Sources* pane, right-click on the **led_test_system** block design and select ***Create HDL Wrapper***.

Select ***Let Vivado manage wrapper and auto-update*** and click ***OK***.

This will create the top-level HDL file for the design.

We must now connect the LEDs_out port of the design to the correct pins on the Zynq device. This is done through the specification of constraints in an XDC file.

(ak) In the *Flow Navigator* window, select ***Add Sources*** from the *Project Manager* section.

The *Add Sources* dialogue will open.

Select ***Add or Create Constraints***, and click ***Next***.

(al) Click the + symbol and then click ***Create File...*** as shown in Figure 4.13.

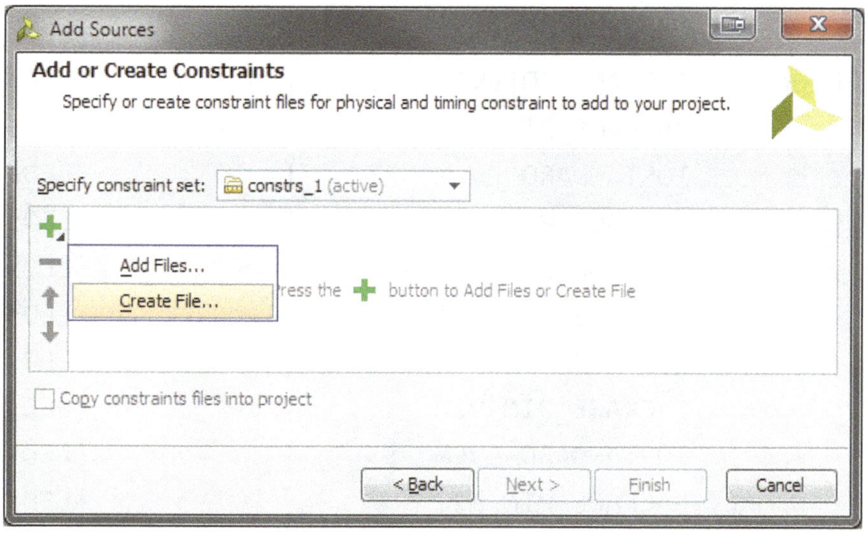

Figure 4.13: Add or Create Constraints Dialogue Window

The *Create Constraints File* dialogue will open.

Select ***XDC*** as the *File type* and enter ***led_constraints*** as the *File name*.

Click ***OK***.

(am)Click ***Finish*** to create the file and close the dialogue.

(an)In the ***Sources*** tab, expand the **Constraints** entry and open the newly created XDC file by double-clicking on ***led_constraints.xdc***.

The file will open in the *Workspace*.

Zed

Add the following lines to the constraints file. Alternatively, they can be copied from the source file available at **C:\Zynq_Book\sources\zedboard\led_controller**.

This connects each individual bit of the LEDs_out port to a specific pin on the Zynq device. The specific pins are connected to the LEDs on the Zedboard.

```
set_property PACKAGE_PIN T22        [ get_ports {LEDs_out[0]}]
set_property IOSTANDARD LVCMOS33    [ get_ports {LEDs_out[0]}]
set_property PACKAGE_PIN T21        [ get_ports {LEDs_out[1]}]
set_property IOSTANDARD LVCMOS33    [ get_ports {LEDs_out[1]}]
set_property PACKAGE_PIN U22        [ get_ports {LEDs_out[2]}]
set_property IOSTANDARD LVCMOS33    [ get_ports {LEDs_out[2]}]
set_property PACKAGE_PIN U21        [ get_ports {LEDs_out[3]}]
set_property IOSTANDARD LVCMOS33    [ get_ports {LEDs_out[3]}]
set_property PACKAGE_PIN V22        [ get_ports {LEDs_out[4]}]
set_property IOSTANDARD LVCMOS33    [ get_ports {LEDs_out[4]}]
set_property PACKAGE_PIN W22        [ get_ports {LEDs_out[5]}]
set_property IOSTANDARD LVCMOS33    [ get_ports {LEDs_out[5]}]
set_property PACKAGE_PIN U19        [ get_ports {LEDs_out[6]}]
set_property IOSTANDARD LVCMOS33    [ get_ports {LEDs_out[6]}]
set_property PACKAGE_PIN U14        [ get_ports {LEDs_out[7]}]
set_property IOSTANDARD LVCMOS33    [ get_ports {LEDs_out[7]}]
```

Zybo Add the following lines to the constraints file. Alternatively, they can be copied from the source file available at **C:\Zynq_Book\sources\zybo\led_controller**:

```
set_property PACKAGE_PIN M14        [ get_ports {LEDs_out[0]}]
set_property IOSTANDARD LVCMOS33    [ get_ports {LEDs_out[0]}]
set_property PACKAGE_PIN M15        [ get_ports {LEDs_out[1]}]
set_property IOSTANDARD LVCMOS33    [ get_ports {LEDs_out[1]}]
set_property PACKAGE_PIN G14        [ get_ports {LEDs_out[2]}]
set_property IOSTANDARD LVCMOS33    [ get_ports {LEDs_out[2]}]
set_property PACKAGE_PIN D18        [ get_ports {LEDs_out[3]}]
set_property IOSTANDARD LVCMOS33    [ get_ports {LEDs_out[3]}]
```

This connects each individual bit of the LEDs_out port to a specific pin on the Zynq device. The specific pins are connected to the LEDs on the Zybo.

Resume (ao) Save the constraints file.

Our simple design is now complete. We can now generate a bitstream.

(ap) In *Flow Navigator*, select ***Generate Bitstream*** from the *Program and Debug* section. If a dialogue window appears prompting you to save your design, click ***Save***.

A dialogue window may open requesting that you launch synthesis and implementation before starting the *Generate Bitstream* process. If it does, click ***Yes*** to accept.

The combination of running the synthesis, implementation and bitstream generation processes back-to-back may take a few minutes, depending on the power of your computer system.

(aq) When bitstream generation is complete a dialogue window will open to inform you that the process as been completed.

Select ***Open Implemented Design***, and Click ***OK***.

With the bitstream generation complete, the final step in Vivado is to export the design to the SDK, where we will create the software application that will allow the Zynq PS to control the LEDs on the Zynq development board.

(ar) Select ***File > Export > Export Hardware...*** from the *Menu Bar*.

The *Export Hardware for SDK* dialogue window will open. Ensure that the option to ***Include bitstream*** is selected, and Click ***OK***.

(as) Launch the SDK in Vivado by selecting ***File > Launch SDK*** from the *Menu Bar* and Click ***OK***.

The SDK will launch.

(at) Once the SDK has launched, create a new **Application Project** by selecting **File > New > Application Project** from the *Menu Bar*.

In the *New Project* dialogue, enter ***LED_Controller_test*** as the *Project name*.

By default the option to create a new board support package will be selected.

Click ***Next***.

(au) In the *Templates* dialogue, select ***Empty Application***, and click ***Finish***.

You should recall that when we created the peripheral in the previous stages of this exercise that a set of software driver files were generated. We must now point the SDK to those driver files. This is done by adding a new repository to the SDK project.

(av) Navigate to **Xilinx Tools > Repositories** in the *Menu Bar*.

In the *Repositories Preferences* window, click on ***New***, as shown in Figure 4.14.

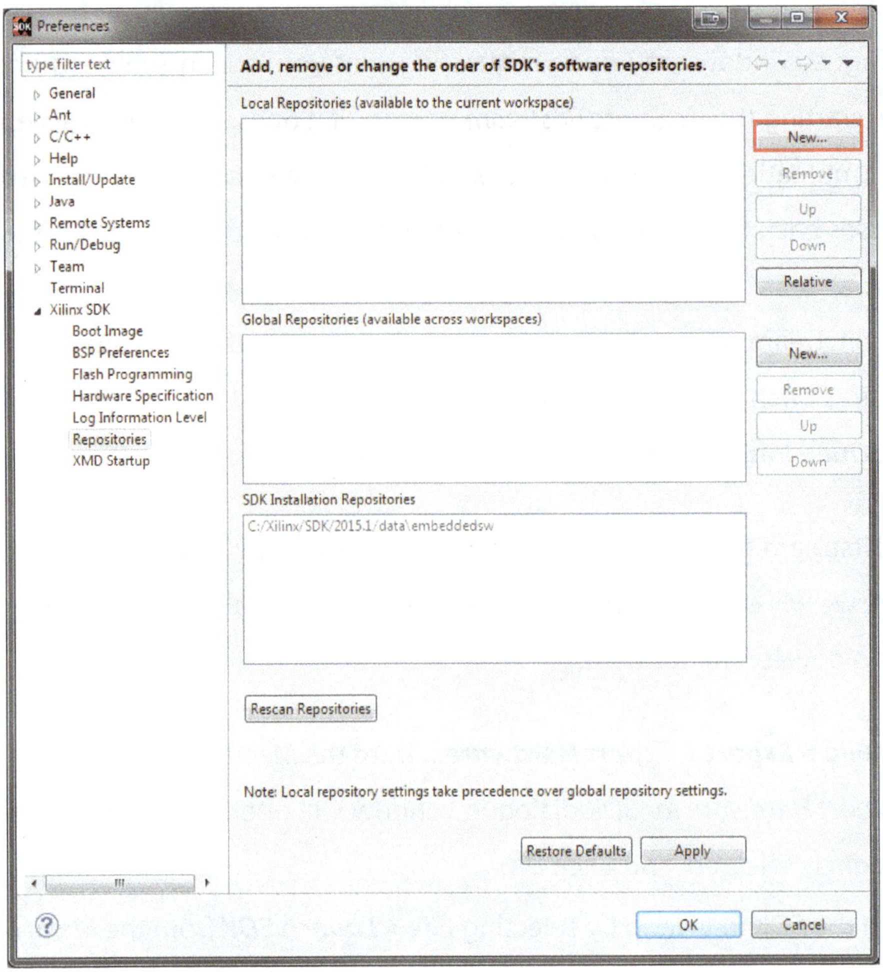

Figure 4.14: SDK Repository Peripherals window

(aw)Browse to the directory:

C:\Zynq_Book\ip_repo\led_controller_1.0

as shown in Figure 4.15, and click ***OK***.

(ax) Close the Repository Preferences window by clicking ***OK***.

Upon closing the preferences window, SDK will automatically scan the repository and rebuild the project to include the driver files.

We must now check that the newly imported driver has been assigned to the LED Controller peripheral.

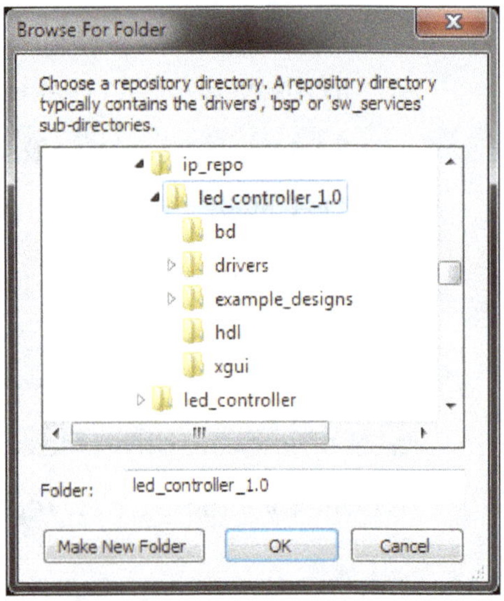

Figure 4.15: led_controller repository selection

(ay) The **system.mss** tab should be open in the Workspace. If it is not, open it by expanding **LED_Controller_test_bsp** in *Project Explorer* and double-clicking on ***system.mss.***

(az) At the top left of the **system.mss** tab, click ***Modify this BSP's Settings***.

The Board Support Package Settings window will open, as in Figure 4.16.

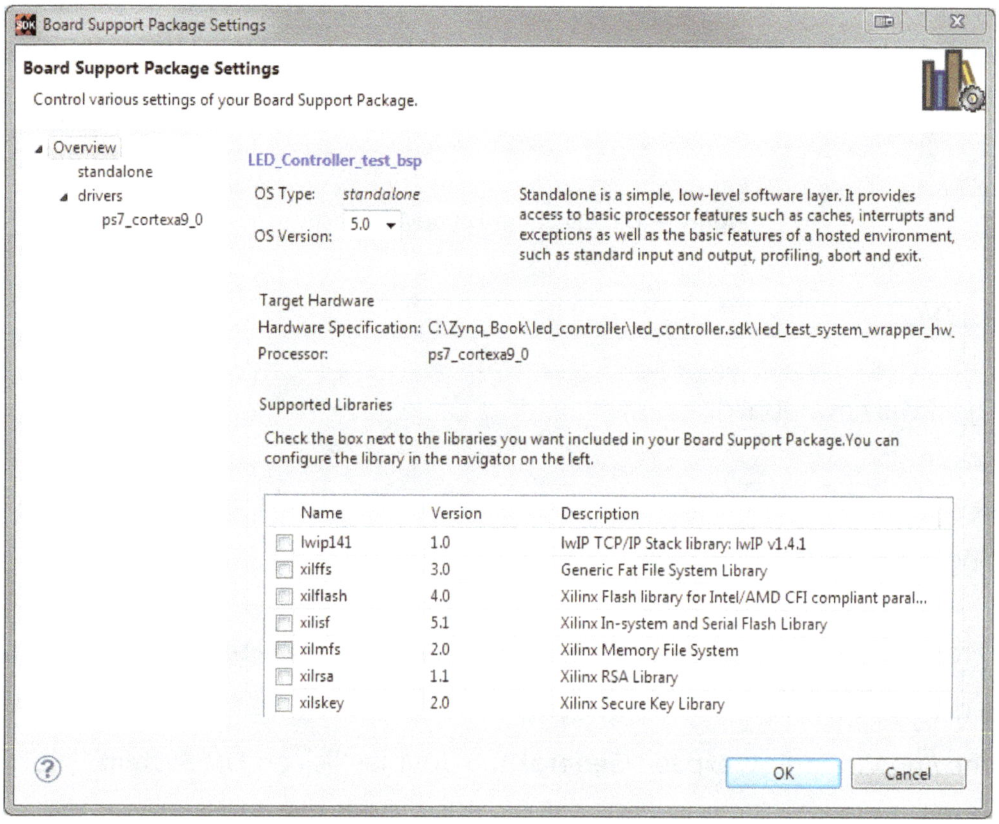

Figure 4.16: Board Support Package Settings window

(ba) Select ***drivers*** from the left-hand menu. From the list of components in the *Drivers* pane, identify **led_controller_0** and ensure ***led_controller*** is selected from the drop-down menu in the **Driver** column, as shown in Figure 4.17.

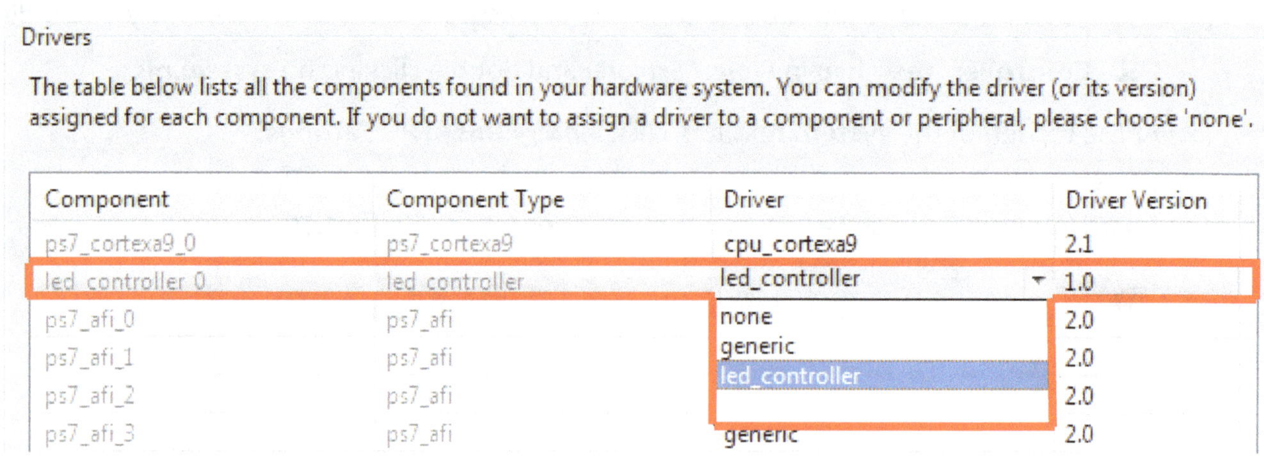

Figure 4.17: LED Controller driver selection

Click ***OK***.

The project will now rebuild.

We can now create a simple C application to control the LEDs. In this instance we will be importing a pre-written source file.

(bb)In *Project Explorer*, expand **LED_Controller_test** and right-click on **src.** Select ***Import*** from the drop-down menu.

In the *Import* window, expand **General** and double-click on ***File System***.

Zed Click ***Browse*** in the top right corner, and navigate to **C:\Zynq_Book\sources\zedboard\led_controller**. Click ***OK***.

In the right-hand panel, select ***led_controller_test_tut_4A.c*** and click ***Finish.***

Zybo Click ***Browse*** in the top right corner, and navigate to **C:\Zynq_Book\sources\zybo\led_controller**. Click ***OK***.

In the right-hand panel, select ***led_controller_test_tut_4A.c*** and click ***Finish.***

Resume The project will rebuild to include the new source file.

Open ***led_controller_test_tut_4A.c*** and examine the functionality.

Before launching the application on the Zynq development board, we must program the Zynq PL and create a new terminal connection.

(bc) From the *Menu Bar*, select **Xilinx Tools > Program FPGA.**

The Bitstream entry should already be populated with the corresponding bitstream that we exported from Vivado earlier.

Click ***Program***, to program the Zynq PL.

NOTE: Once the device has successfully been programmed, the *DONE LED* on the ZedBoard will turn blue. Similarly the *DONE LED* on the Zybo will turn green.

(bd) Select the ***Terminal*** tab from the *Console* window at the bottom of the workspace, as in Figure 4.18.

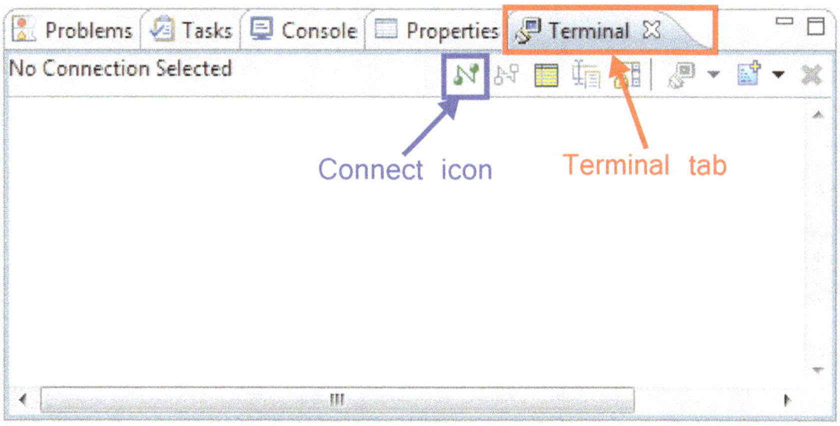

Figure 4.18: SDK Terminal tab

(be) Click the Connect icon (as highlighted in Figure 4.18).

(bf) The Terminal Settings window will open. Configure the settings as specified in Figure 4.19.

NOTE: The value of the *Port* entry will vary depending on which the USB UART cable is connected to.

In order to determine this value on a Windows system, open the Device Manager and identify the COM port (may be named 'USB Serial Port').

(bg) Click OK to initiate the new Terminal connection.

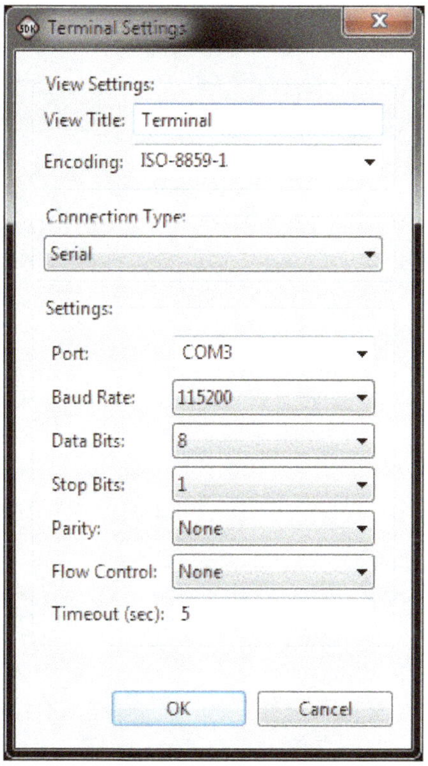

Figure 4.19: Terminal Settings

Now that the Zynq PL is programmed, and the Terminal connection has been created, we can program the Zynq PS with our software application.

(bh)In Project Explorer, right-click on **LED_Controller_test** and select ***Run As > Launch on Hardware (GDB)***, as shown in Figure 4.20.

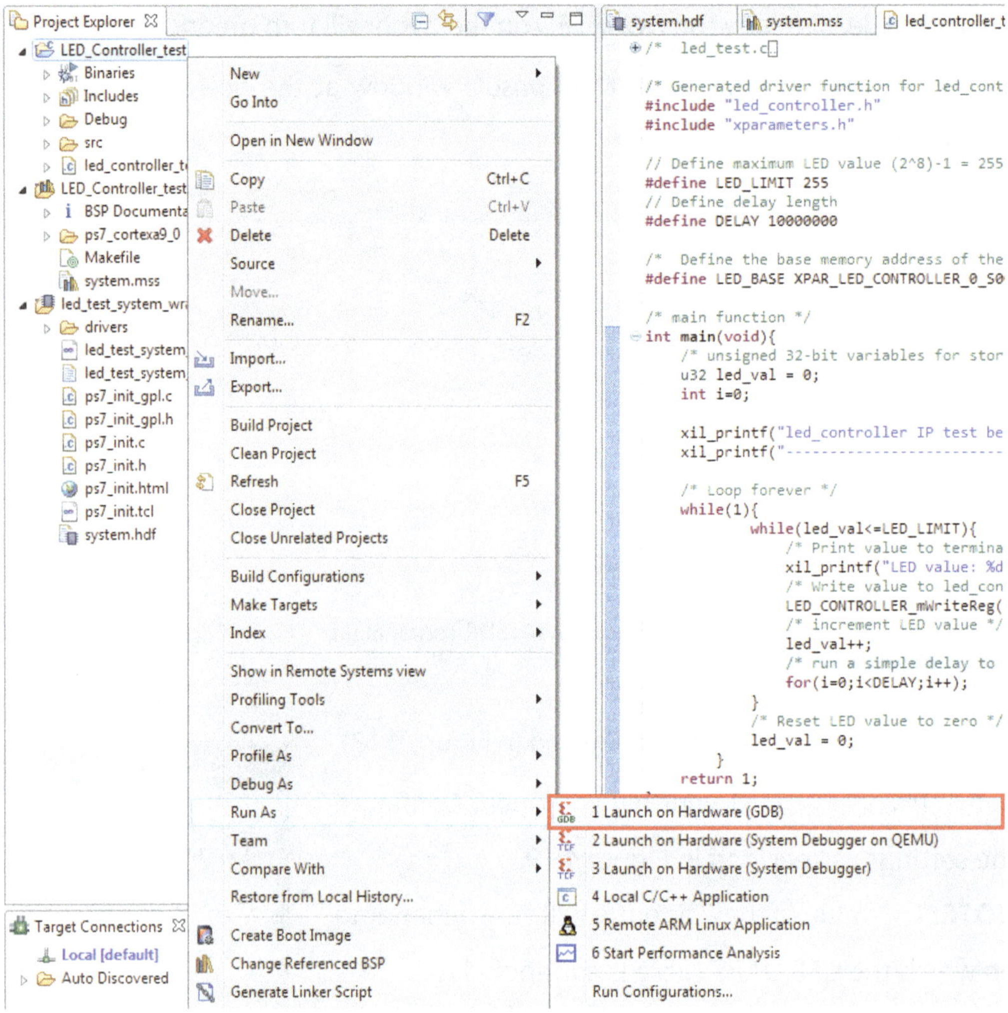

Figure 4.20: Run Application on hardware

(bi) Switch to the Terminal tab of the Console window, and confirm that the LED value is being output, as in Figure 4.21.

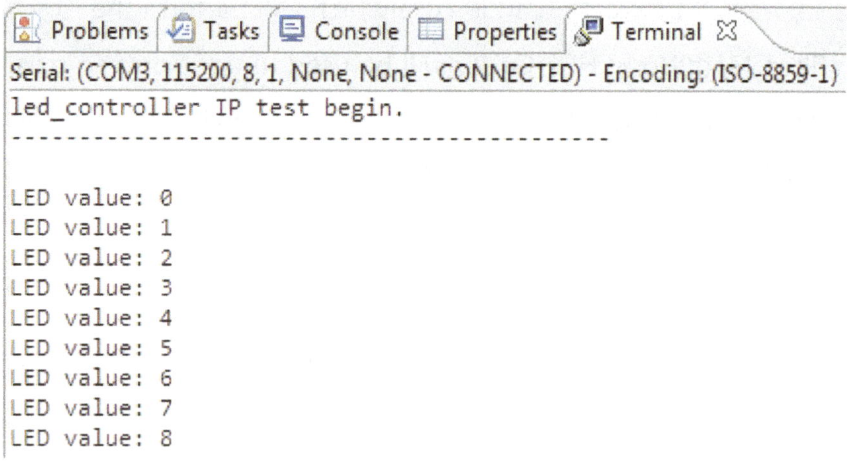

Figure 4.21: Terminal tab displaying LED values

You should also see the LEDs on the development board displaying the corresponding LED values.

This concludes this exercise on designing Zynq IP in HDL. You should now be familiar with:

- Creating AXI interface templates with the Create and Package IP Wizard.
- Adding functionality to HDL IP peripherals in Vivado and IP Packager.
- How to connect packaged IP to a Zynq Processing System in IP Integrator.
- Creating software applications to control the HDL IP using the generated C software drivers, and executing them on a Zynq development board

Exercise 4B Creating IP in MathWorks HDL Coder

In this exercise, we will be creating an IP core which will perform the function of an LMS noise cancellation filter. MathWorks HDL Coder will be used to transform an existing Simulink block-based model into an RTL description which will be packaged for use in the Vivado IP Catalog. We will start by opening the Simulink model in MATLAB.

Before starting this exercise, you should copy some source files into a new working directory.

(a) In Windows Explorer, navigate to **C:\Zynq_Book\sources\hdl_coder_lms** and copy the contents of the directory to a new directory called **C:\Zynq_Book\hdl_coder_lms**.

(b) Launch MATLAB by navigating to **Start** > **All Programs** > **MATLAB** > **R2015a** > **MATLAB R2015a**

MATLAB will open and you will see the main workspace, as shown in Figure 4.22 (or a variation thereof).

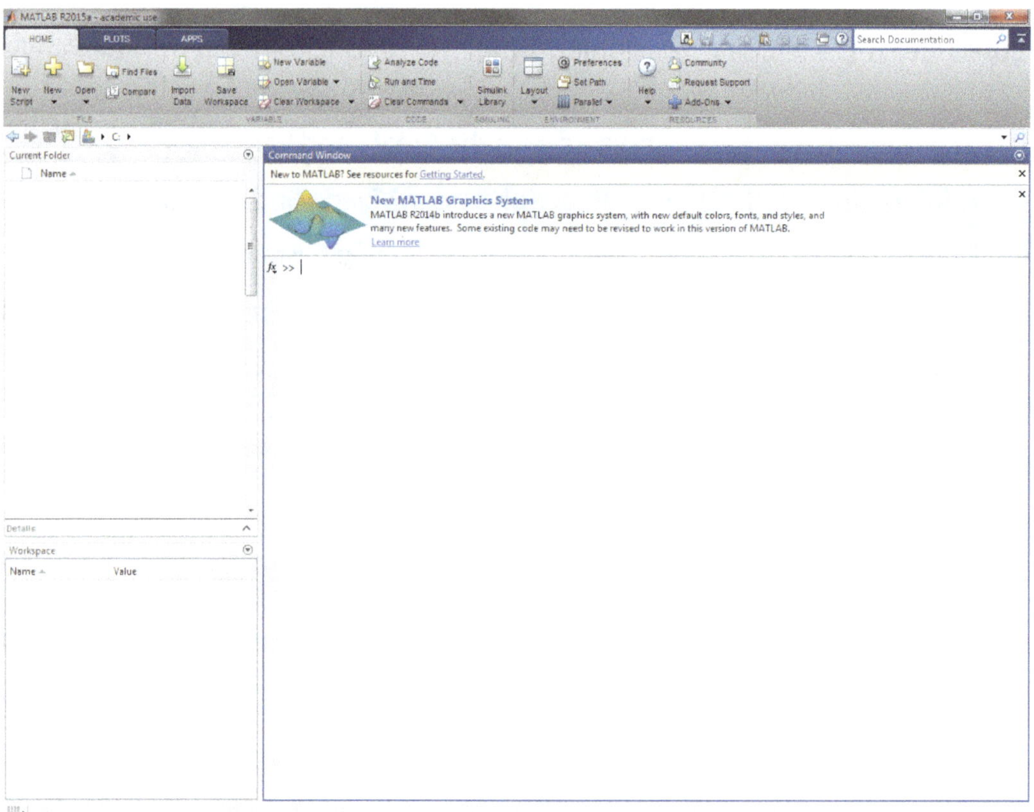

Figure 4.22: MATLAB workspace environment

Note: This workbook uses version R2015a of MATLAB. If you have a different MATLAB version you may need to replace your own version (i.e. R2014a/R2014b/R2013a/R2013b) with 2015a.

(c) If your MATLAB **HDL Toolpath** has already been set-up, move on to Step (d), otherwise carry-out the following procedure.

- Download and install **Xilinx ISE 14.7** from the Xilinx website or from the following link: http://www.xilinx.com/products/design-tools/ise-design-suite.html
- Using *Windows Explorer* locate the *Xilinx ISE Application* within the Xilinx ISE 14.7 installation directory named **ise.exe**. Typically, the application can be found at the following address if installed to the C Drive:
  C:\Xilinx\14.7\ISE_DS\ISE\bin\nt\ise.exe
- Copy the address to the clipboard and open the MATLAB workspace shown previously in Figure 4.22.
- In the **Command Window** enter the following function:

```
hdlsetuptoolpath('ToolName', 'Xilinx ISE', 'ToolPath',
'C:\Xilinx\14.7\ISE_DS\ISE\bin\nt\ise.exe')
```

  Where the fourth parameter (application address) is the installation directory previously copied to the clipboard.
- Successfully setting up the **HDL Toolpath** will result in the following information being displayed:

```
>> hdlsetuptoolpath('ToolName', 'Xilinx ISE', 'ToolPath', 'C:\Xilinx\14.7\ISE_DS\ISE\bin\nt\ise.exe')
Setting XILINX environment variable to:
C:\Xilinx\14.7\ISE_DS\ISE
Setting XILINX_EDK environment variable to:
C:\Xilinx\14.7\ISE_DS\EDK
Setting XILINX_PLANAHEAD environment variable to:
C:\Xilinx\14.7\ISE_DS\PlanAhead
```

- HDL Coder can now be used to synthesise HDL code for Xilinx Hardware Platforms.

(d) Enter **C:\Zynq_Book\hdl_coder_lms** as the working directory, as highlighted in Figure 4.23.

Figure 4.23: Setting the MATLAB working directory

In the *Current Folder* pane, you should also see four files:

- **original_speech.wav** — A short audio clip of speech.
- **setup.m** — Performs setup commands to import the audio samples into the MatLab workspace and set the system sample rate accordingly.
- **lms.slx** — A simulink model which implements and LMS noise cancellation process.
- **playback.m** —Can be used to verify the LMS filtering process via audio playback of the various stages.

The setup commands in **setup.m** are automatically called when the Simulink simulation is initialised.

(e) Open the LMS Simulink model by double-clicking on ***lms.slx*** in *Current Folder* pane. The model should open and you should see the LMS system, as shown in Figure 4.24.

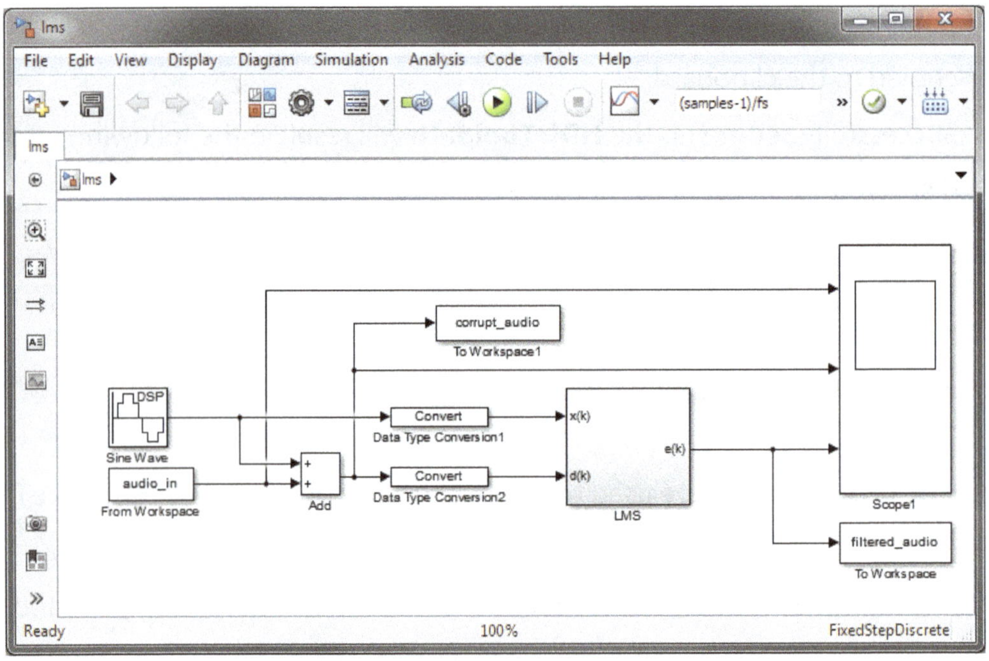

Figure 4.24: LMS model in Simulink

The model features two sources:

- a ***Sine Wave*** block which generates tonal noise.
- A ***From Workspace*** block which imports the audio samples from the MATLAB Workspace.

The tonal noise is then added to the audio samples to create a corrupted audio signal.

In order to generate HDL code for the Simulink LMS model using HDL Coder, the inputs to the system must be in fixed-point numerical format. Two ***Data Type Conversion*** blocks are used to convert the corrupt audio signal and the tonal noise signal to fixed-point format. The fixed-point signals are then input to an LMS subsystem, which we will explore in the next step.

At the output of the LMS subsystem, the error signal, **e(k)**, is input to a scope along with the corrupt audio and tonal noise inputs, for visual inspection of the signals. Two ***To Workspace*** blocks are also present to allow the LMS output and the corrupt audio signals to be output to the MatLab workspace for audio playback.

(f) Drill down into the LMS subsystem block by double-clicking on it. You will see the system in Figure 4.25.

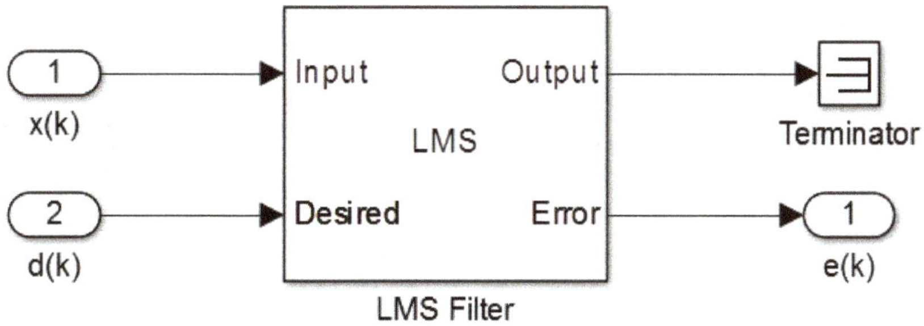

Figure 4.25: LMS subsystem

It features a single ***LMS Filter*** block. As we are not interested in the ***Output*** signal, it is unconnected. Further reading about the functionality of an *LMS Filter* can be found by ***right clicking*** the *LMS Filter block* and selecting ***Help*** as shown below:

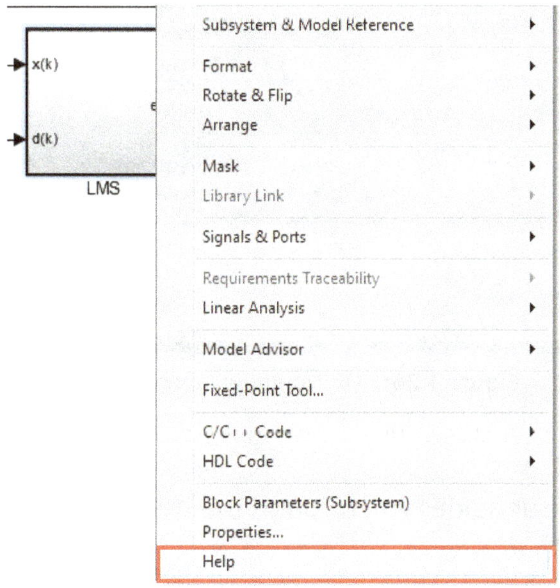

(g) Open the *LMS Filter Block Parameters* by double-clicking on the ***LMS Filter*** block.

Take a moment to explore the parameters. You should be able to determine that there are 16 ***adaptive filter coefficients*** and a ***step size*** of 0.125.

(h) Close the Parameters window, and return to the main Simulink model by clicking the ***Up To Parent*** button,  .

We will be generating HDL code for the LMS subsystem only.

Right-click on the LMS subsystem and select **HDL Code > HDL Workflow Advisor**.

The HDL Workflow Advisor window will open, as in Figure 4.26.

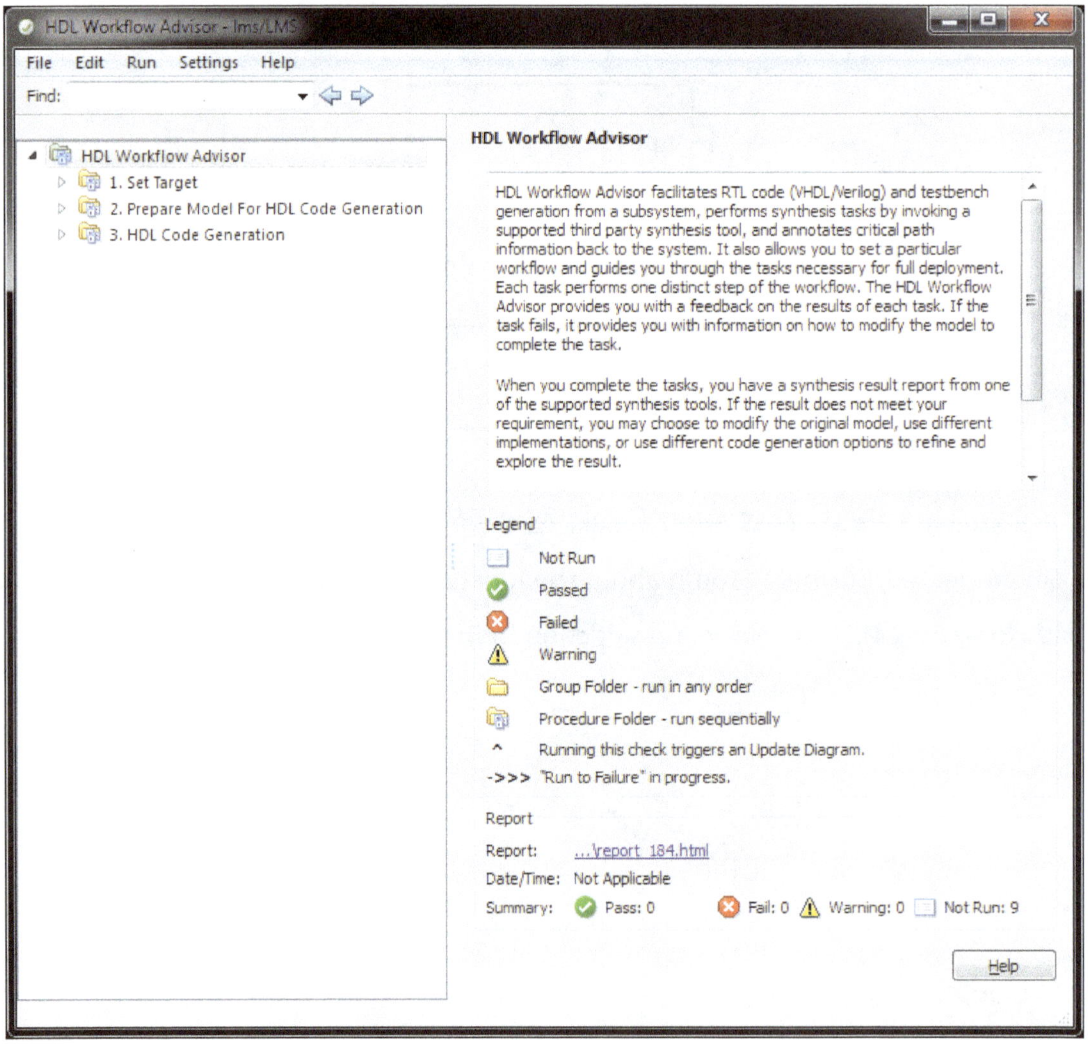

Figure 4.26: HDL Workflow Advisor window

The HDL Workflow Advisor guides you through the steps required to generate RTL code for your design.

(i) In the left-hand panel, expand ***Set Target*** and select ***1.1. Set Target Device and Synthesis Tool***.

Here we specify the output format of the RTL and the target platform.

(j) In the *Input Parameters* pane, select ***IP Core Generation*** as the *Target workflow*, and ***Generic Xilinx Platform*** as the *Target platform*

Zed

At this stage, additional part specification options will now be available. Target the **Zedboard** by first confirming the required part by inspecting the Zynq chip on the board. Enter the part details into HDL Coder as in Figure 4.27.

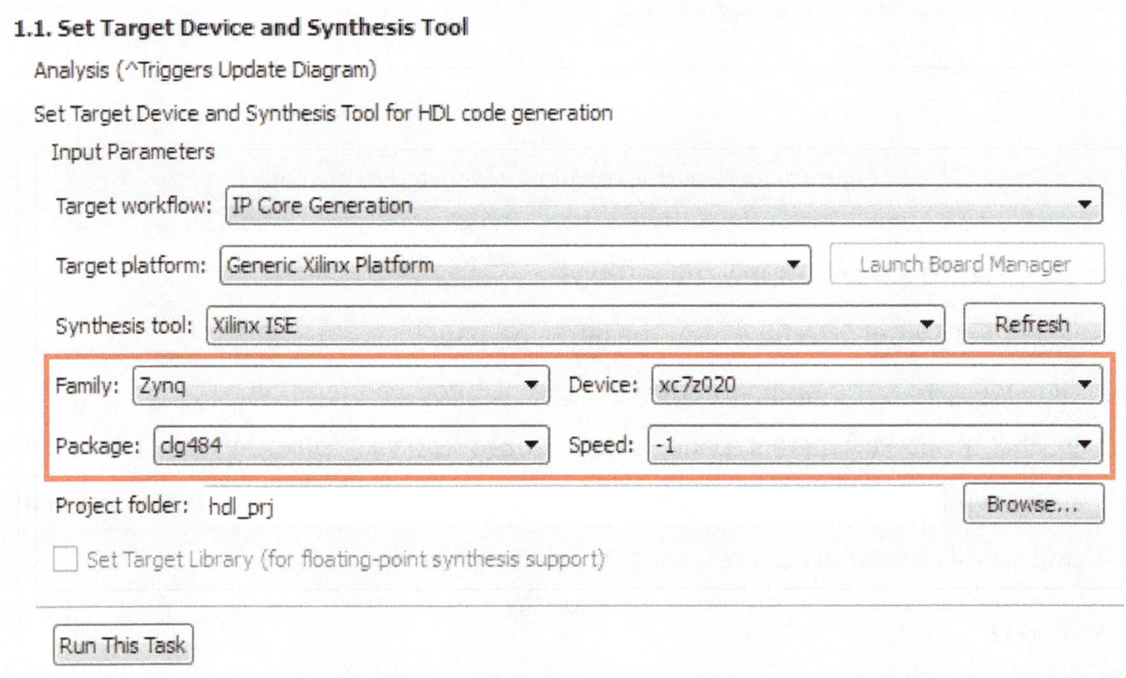

Figure 4.27: ZedBoard HDL Workflow Advisor Input Parameters

Zybo

At this stage, additional part specification options will now be available. Target the **Zybo** by first confirming the required part by inspecting the Zynq chip on the board. Enter the part details into HDL Coder as in Figure 4.28.

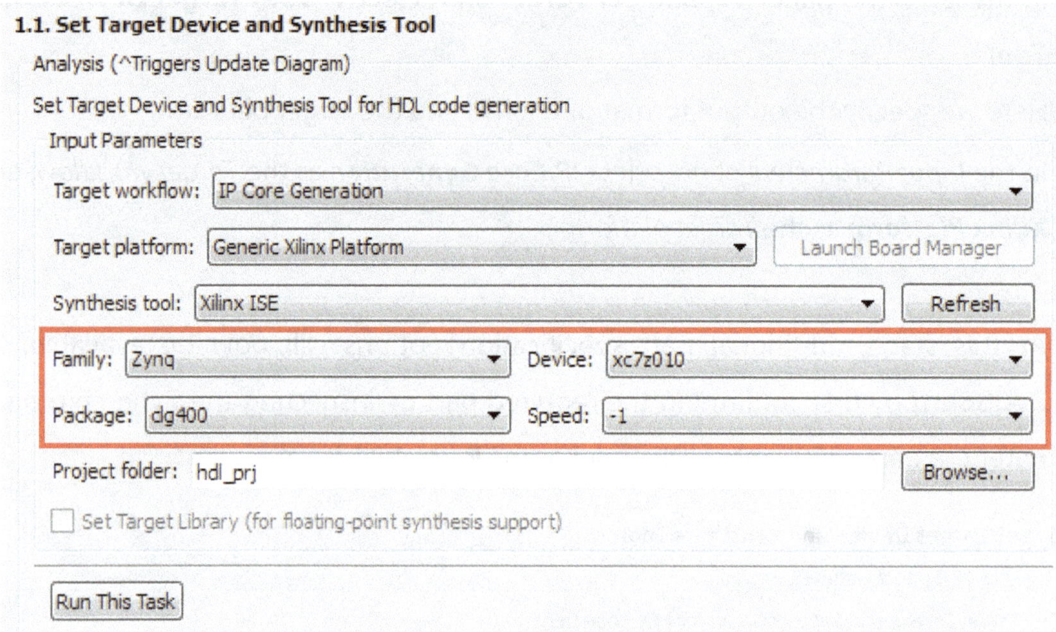

Figure 4.28: Zybo HDL Workflow Advisor Input Parameters

Resume (k) Click ***Run This Task*** to apply the settings.

(l) Select ***Set Target Interface*** from the left hand panel.

Here we specify the target interface for the HDL code generation.

In the Input Parameters pane, select ***Coprocessing - blocking*** as the *Processor/FPGA synchronization*. This will automatically infer an **AXI4-Lite** interface for all ports in the design, and specify a memory address for each as shown in Figure 4.29.

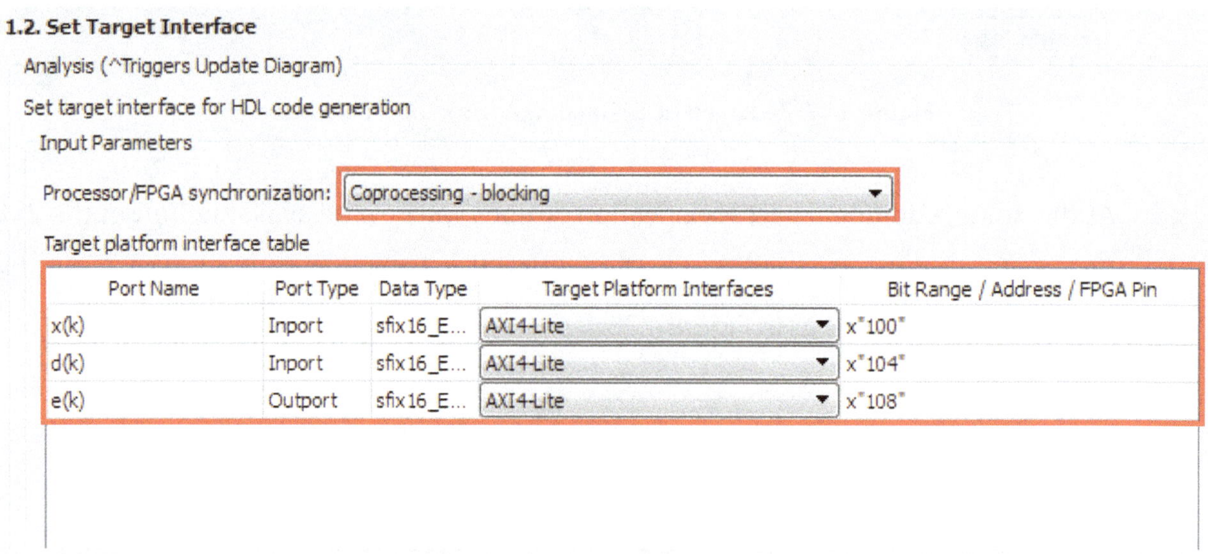

Figure 4.29: HDL Workflow Advisor Set Target Interface

(m) Click ***Run This Task*** to apply the settings.

(n) Expand ***Prepare Model for HDL Code Generation*** in the left hand panel, and select ***Check Global Settings***.

Here, model-level settings will be checked to verify if the model is ready for HDL code generation.

(o) Click ***Run This Task*** to check the model-level settings.

If this step fails, click ***Modify All*** to allow *HDL Workflow Advisor* to modify the settings. This step should now pass, and you will be presented with a table of the results.

The next few steps are all checks, and can be performed in batch.

(p) Right-click on *Check Sample Times* in the left hand pane, and select ***Run to Selected Task*** as shown in Figure 4.30.

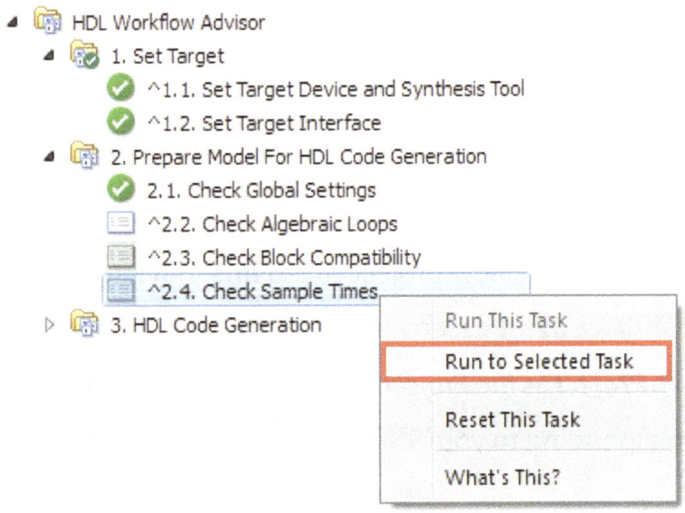

Figure 4.30: HDL Workflow Advisor Run to Selected Task

(q) This will perform the checks one after another to prevent you from running each individually. All checks should pass.

The final steps involve specifying basic settings about the RTL code, such as what language to use (VHDL/Verilog), and what code generation reports to generate. Finally the HDL code will be generated.

(r) Expand *HDL Code Generation* in the left hand pane, and further expand ***Set Code Generation Options***.
Click on ***Set Basic Options***.

(s) Select ***VHDL*** as the **Language** in the *Target* pane.

You can also select any of the *Code generation reports* that you would like.

(t) Select ***Set Advanced Options*** in the left hand panel.

Here you can specify more advanced options for the HDL code.

We will be leaving the values as default, but you may wish to explore the settings for future use.

(u) Right-click on *Set Advanced Options*, and select ***Run to Selected Task*** to apply the settings.

(v) Finally, select *Generate RTL Code and IP Core* from the left hand panel.

This is the step which will finally generate the HDL code for the LMS IP Core.

Set the *IP core name* as ***lms_pcore*** and click ***Run This Task***.

Once HDL Coder has finished generating the HDL code, the Code Generation Report window will open. This provides a summary of the HDL Coder results and provides further information on the target interface and clocking.

The final stage of creating our LMS IP core is to package it with IP Packager so that we can use it in IP Integrator designs. To do this we will need to create a new Vivado project.

(w) Launch Vivado and create a new project called **lms_packaging** at the following location: **C:\Zynq_Book\hdl_coder_lms**, ensuring that the option to ***create a project subdirectory*** is selected. Set ***RTL Project*** as the *Project Type*, select ***VHDL*** as the *target language*, and enter the *default part* corresponding to your Zynq development board.

For more detail on the process of creating a new Vivado project, refer to **Step (a)** of **Exercise 4A**.

(x) When the project has been created and opened, select ***Tools > Create and Package IP*** from the *menu bar*, and Click ***Next***.

(y) Select the option to ***Package a specified directory***, and click ***Next***.

(z) Enter **C:/Zynq_Book/hdl_coder_lms/hdl_prj/ipcore/lms_pcore_v1_00_a** as the *IP Location*.

(aa) Click ***Next*** to move to the *Edit in IP Packager Project Name* dialogue, and click Next to accept the default *Project Name* and *Project Location*.

(ab) At the *Summary* window, and click ***Finish*** to launch **IP Packager**.

(ac) In the left hand panel of the *IP Packager* window, select ***Ports and Interfaces***.

The *IP Interfaces* panel will open, and you should see that **IP Packager** has identified the individual AXI ports, but has not inferred an AXI interface.

To infer an AXI interface:

(ad) Right-click on a blank section of the *IP Ports and Interfaces* pane, and select ***Auto Infer Interface ...***

(ae) The *Auto Infer Interface Chooser* window will open:

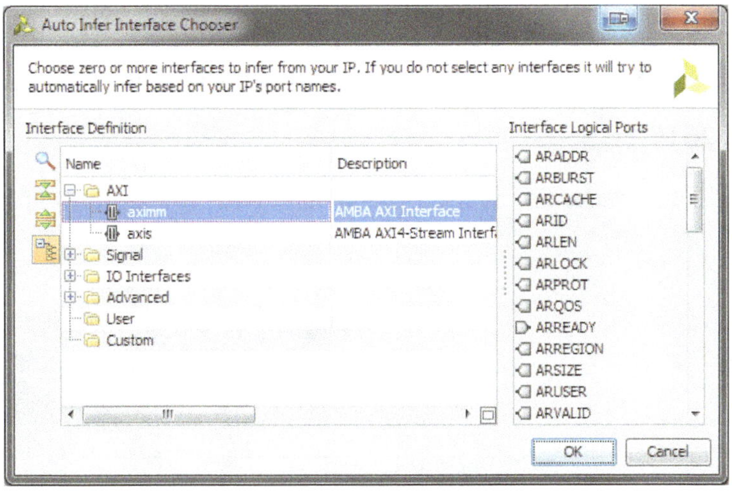

Select ***aximm*** from the list, as shown, and click ***OK***.

The individual AXI ports in our design will be mapped to an **AXILite** interface.

(af) Select ***Addressing and Memory*** from the left hand panel. Here, IP Packager has incorrectly specified an address **Range** of **65536**. Click on the **Range**, and change the value to ***32***.

(ag) Finally, select ***Review and Package*** from the left hand menu.

Review the information provided, and click ***Package IP***.

This completes the generation of an LMS component from Mathworks HDL Coder. You should now be familiar with:

- Using the Simulink block-based design environment for the design and simulation of IP.
- Using the HDL Workflow Advisor to guide you through the steps of generating RTL code and IP cores for existing Simulink designs.
- Packaging HDL Coder generated IP blocks in IP Packager for use in Vivado IP Integrator designs.

Exercise 4C Creating IP in Vivado HLS

In this final exercise, we will creating an IP core that will implement the functionality of an NCO. The tool that we will be using is Vivado HLS, and we shall explore some of the features which allow us to specify arbitrary precision fixed-point data types, as well as the directives required to export IP with an AXI-Lite slave interface, to allow the IP core to interface with the Zynq processor.

We will start by creating a new project in Vivado HLS.

(a) Launch Vivado HLS by double-clicking on the Vivado HLS desktop icon:  , or by navigating to **Start > All Programs > Xilinx Design Tools > Vivado 2015.1 > Vivado HLS > Vivado HLS 2015.1**

(b) When Vivado HLS loads, you will be presented with the *Getting Started* screen, as in Figure 4.31.

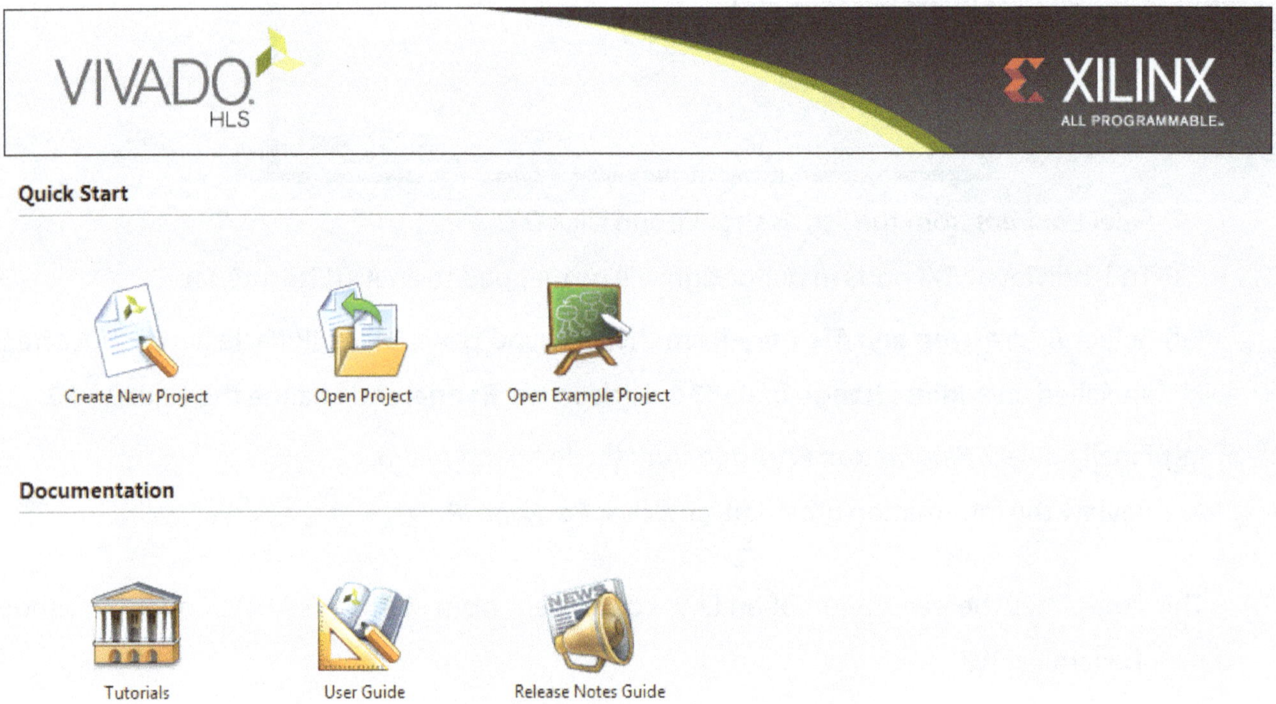

Figure 4.31: Vivado HLS Getting Started screen

(c) Select the option to ***Create New Project*** and the *New Vivado HLS Project Wizard* will open, as in Figure 4.32.

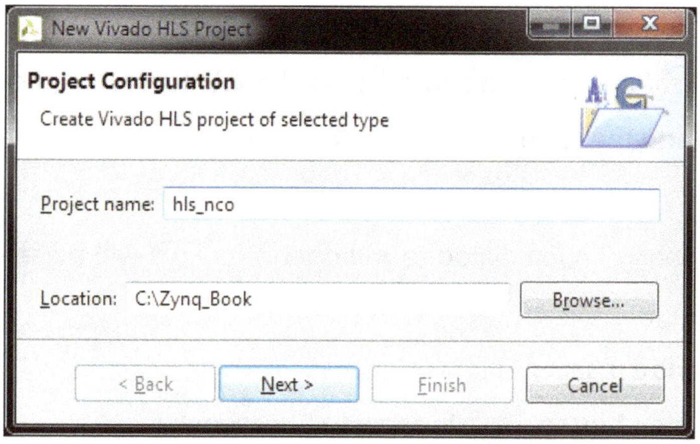

Figure 4.32: Vivado HLS New Project Wizard

Enter ***hls_nco*** as the *Project name*, and ***C:\Zynq_Book*** as *Location*.

Ensure that the options match those in Figure 4.32, and click ***Next***.

(d) The *Add/Remove Files* dialogue will appear. This is where existing C-based source files can be added to the project, or new files created.

Enter ***nco*** as the *Top Function* and click ***Add Files...***

Navigate to ***C:\Zynq_Book\sources\hls_nco*** and select ***nco.cpp***. Click ***Open***.

The dialogue should now resemble Figure 4.33.

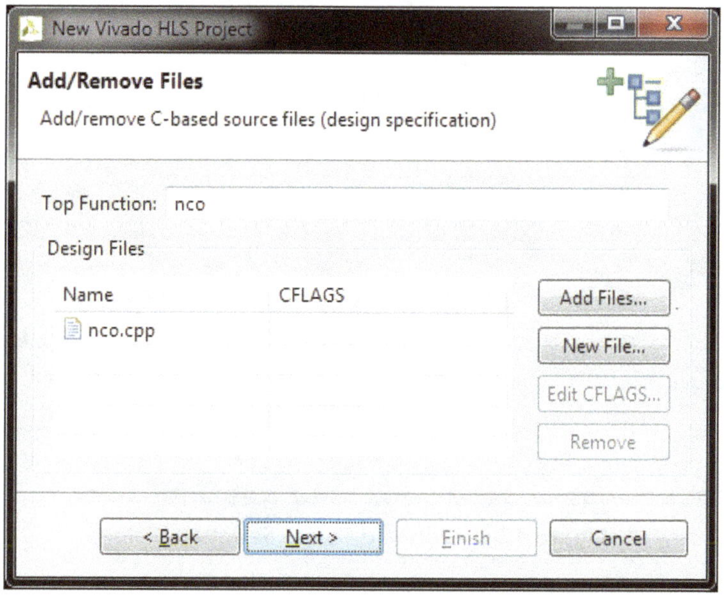

Figure 4.33: Vivado HLS New Project Wizard (Add/Remove Files)

Click ***Next***.

(e) A second *Add/Remove Files* dialogue will appear. This is where C-based testbench files can be added to the project, or new files created.

Click ***Add Files...*** and navigate to ***C:\Zynq_Book\sources\hls_nco***. Select ***nco_tb.cpp*** and click ***Open*** to add the testbench file to the project.

Click ***Next***.

(f) The *Solution Configuration* dialogue will open. Here we will be selecting the part which we will be targeting.

Ensure the **Period** is set to 10.

Click the selection button, [...], in the *Part Selection* pane.

The *Device Selection Dialog* will open.

Zed As we are targeting the ZedBoard, select ***Boards*** in the *Specify* pane and choose ***ZedBoard Zynq Evaluation and Development Kit***, as in Figure 4.34.

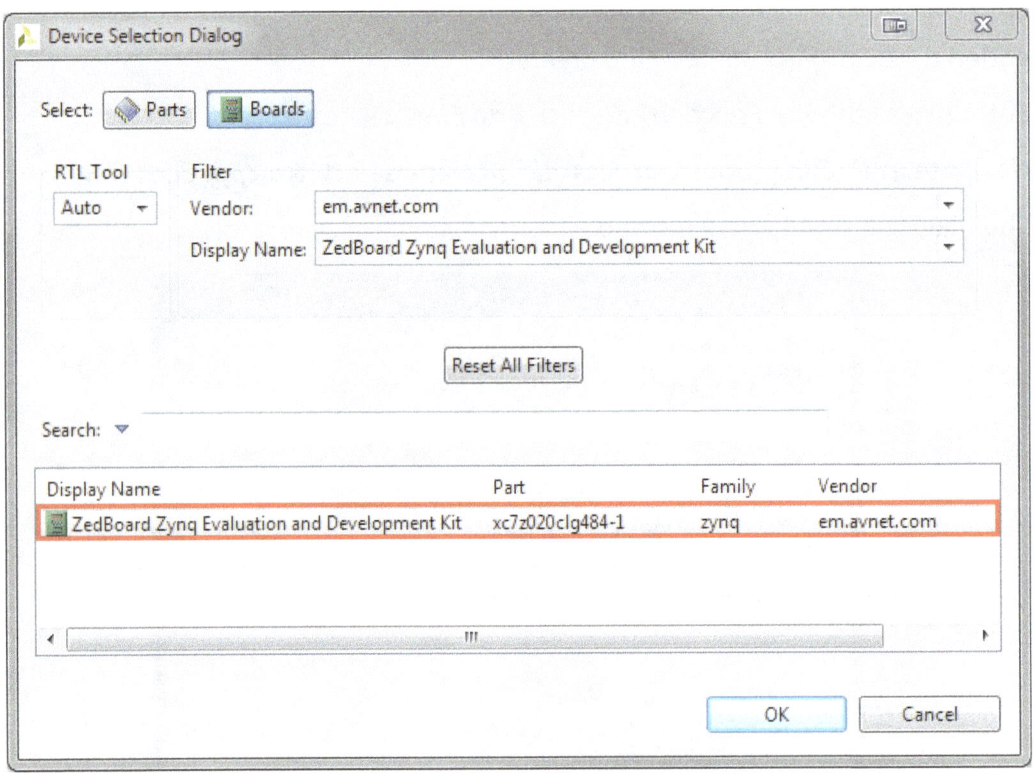

Figure 4.34: Zedboard Device Selection Dialog

Click ***OK*** to close the dialogue and return to the *New Project Wizard*.

Zybo As we are targeting the Zybo, select ***Parts*** in the *Specify* pane and then filter the board parts using the filter drop down menus, as shown in Figure 4.35.

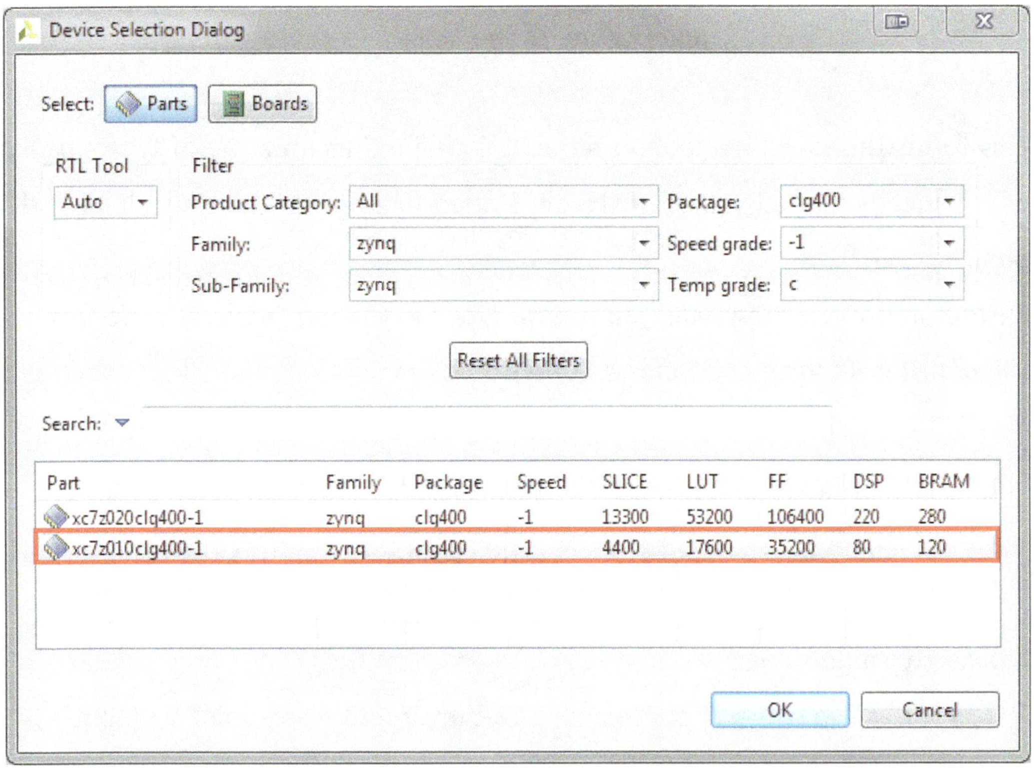

Figure 4.35: Zybo Part selection dialogue

The required part can be confirmed by inspecting the Zynq chip on the Zybo development board. The **Z7010** Zynq chip with a **clg400** package should be selected. Click ***OK***

Resume (g) Click ***Finish*** to close the *New Project Wizard* and to create the project.

The Vivado HLS workspace will open.

(h) In the *Explorer* panel, expand the ***Source*** and ***Test Bench*** headings. You should see the source files that we specified in the *New Project Wizard*, as in Figure 4.36.

(i) Open ***nco.cpp*** and examine the contents of the file.

You should notice the inclusion of the header file ***ap_fixed.h*** on the first line. This is the arbitrary precision fixed-point library which adds support for the use of fixed-point data types in C++.

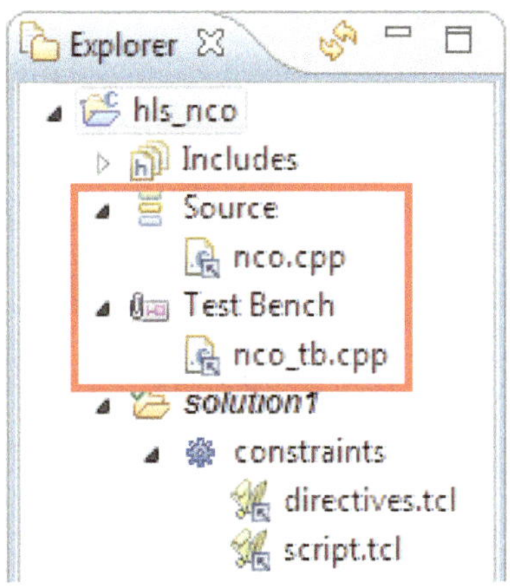

Figure 4.36: Vivado HLS Explorer panel

The next thing that you should see is the global declaration of a 2^{12} = 4096 value array:

```
const ap_fixed<16,2> sine_lut[4096] ...
```

This forms the sinewave lookup table. It is defined as an array of type `ap_fixed<16,2>`, which means that all values are16-bit, signed fixed-point (2 integer bits and 14 fractional bits).

Further information on fixed-point data types in Vivado HLS can be found in ***Chapter15 - Vivado HLS: A Closer Look*** of the **Zynq Book.**

The functionality of the NCO is contained in the function:

```
void nco (ap_fixed<16,2> *sine_sample, ap_ufixed<16,12> step_size)
```

It takes two arguments:

- **`*sine_sample`** — A pointer to a 16-bit, signed fixed-point variable which forms the output sample of the NCO.
- **`step_size`** — 16-bit, unsigned fixed-point value which provides the step size input for the NCO.

(j) Explore the **nco** function, ensuring that you understand it all.

Open ***nco_tb.cpp***. This is the testbench file which is used to ensure that the functionality of the C-based source file is correct.

Explore the code in the file, ensuring that you understand the functionality.

This is a simple file which opens a text file in write-mode, to allow you to output the sinusoidal samples. It then calls the **nco** function from within a *for-loop* in order to generate a finite number of sinusoidal samples, which are then output to the text file.

The text file is formatted in a way which easily allows you to import the samples into MATLAB for analysis.

Note: The location of the output file is determined by the following line in the testbench file:

```
char *outfile = "C:\\Zynq_Book\\nco_sine.m";
```

You should change the output file path accordingly to a location on your local machine.

We will now run a C simulation.

(k) Click the ***Run C Simulation*** button, , from the *Main Toolbar*.

The *C Simulation Dialog* window will open. Click ***OK*** to run the simulation with the default settings.

The C simulation will run, and you should see the following output in the Console window:

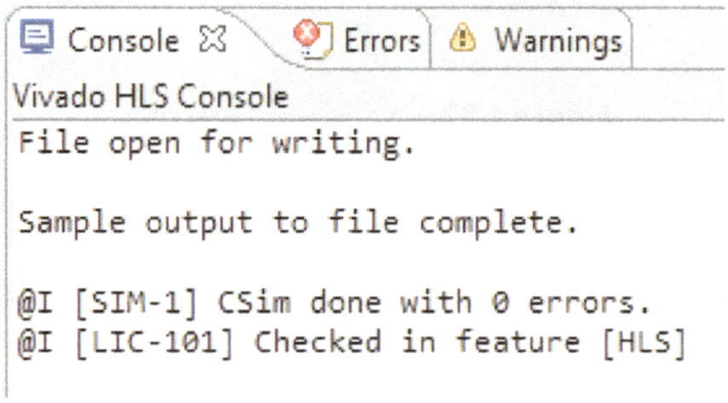

The sine wave samples that were generated by the NCO will have been output to the location which you specified in the previous step.

If you wish, you can import the sine wave samples into MATLAB using the output file to verify that the NCO has correctly generated a sine wave. This should be done at your own discretion, and will not be covered in this exercise.

The process of HLS has been covered previously in **The Zynq Book Tutorial: Designing With Vivado High Level Synthesis**, and you should refer to it for more detailed information on the various steps involved. For the purposes of this exercise, it is presumed that you have a reasonable knowledge of the Vivado HLS tool.

As we want to allow our NCO peripheral to be controlled by a Zynq PS, it is necessary to give it an interface. This can be achieved using a variety of interfaces such as the AXI interface or a GPIO for simple data transfers. The AXI interface will be used; this is carried out in Vivado HLS through the use of directives.

(l) Ensure that **nco.cpp** is the active source file, and select the ***Directive*** tab in the right-hand side of the Vivado HLS workspace, as shown in Figure 4.37.

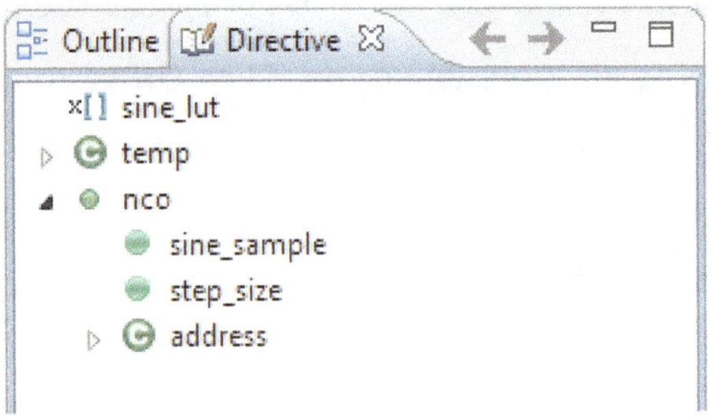

Figure 4.37: Vivado HLS Directive tab

First, we will define the interface of the NCO as an **AXI-Lite slave**.

(m) Right-click on ***nco*** in the *Directive* tab, and select ***Insert Directive***.
As the *Directive Type*, select ***INTERFACE***.
Leave *Destination* as ***Directive File.***
Select ***s_axilite*** from the mode drop-down menu.
Click ***OK***.

We will now define the NCO as having a **ap_ctrl_none** interface, to remove unneeded control signals.

(n) Right-click on ***nco*** in the *Directive* tab, and select ***Insert Directive***.
As the *Directive Type*, select ***INTERFACE***.
Leave *Destination* as ***Directive File.***
Select ***ap_ctrl_none*** from the *mode* drop-down menu.
Click ***OK***.

Finally, we will be defining the two variables, **sine_sample** and **step_size**, as ports on the **AXI-Lite slave** interface.

(o) Right-click on ***sine_sample*** in the *Directive* tab, and select ***Insert Directive***.
As the *Directive Type*, select ***INTERFACE***.
Leave *Destination* as ***Directive File.***
Select ***s_axilite*** from the mode drop-down menu.
Click ***OK***.

(p) Repeat the previous step for the **step_size** variable in the *Directive* tab.

On completion, the Directive tab should look like Figure 4.38.

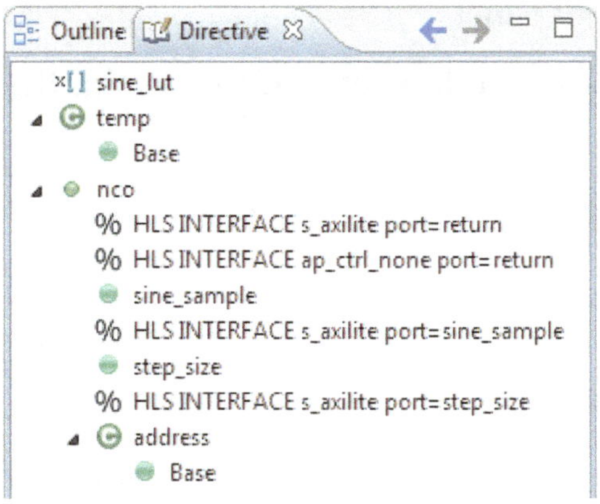

Figure 4.38: Complete Directive tab

We can now run HLS.

(q) Run C Synthesis by clicking the ***Run C Synthesis*** button, ▷, from the *Main Toolbar*.

(r) Click the Export RTL button, ⊞, from the Main Toolbar.

The Export RTL Dialog window will open, as shown in Figure 4.39.

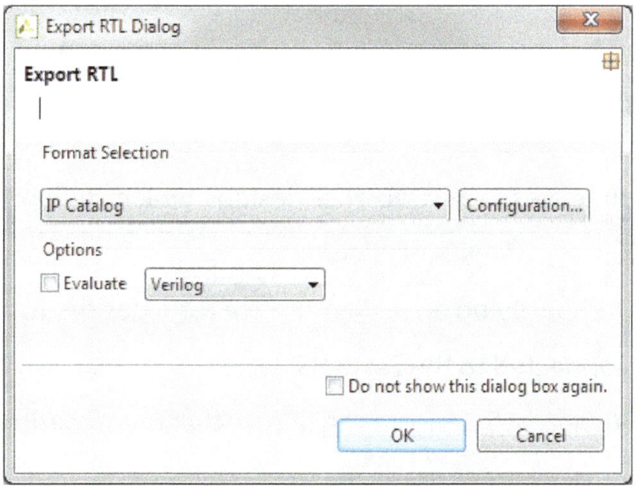

Figure 4.39: Vivado HLS Export RTL Dialog Window

(s) Select **IP Catalog** as the *Format Selection*.

If you choose, you can edit the *IP Identification* data by clicking the ***Configuration*** button. Additionally, the IP core can be generated using Verilog or VHDL. Vivado is capable of synthesising mixed hardware languages. We will keep the default option using Verilog.

(t) Click ***OK*** to generate the IP core.

When RTL Generation has completed, a directory named ***impl*** will be visible in the *Explorer* panel
This directory contains the ***ip*** subdirectory which contains the generated IP package.
Take a moment to explore the contents of the ***ip*** directory.

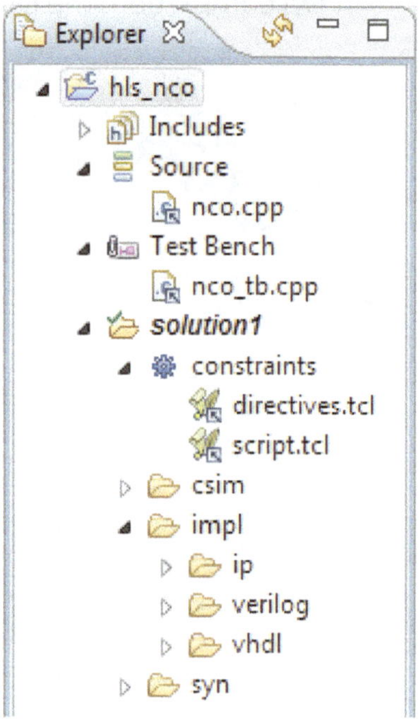

With the IP generated, the next step would be to include it in an IP Integrator design (which will be covered in the next tutorial). For future reference, however, it is worth briefly describing how this would be done.
In order to include HLS generated IP in IP Integrator, it must first be added to the Vivado IP Catalog. To do this you must add the output from HLS to an IP repository. This can be achieved by either adding the HLS generated output directory to an existing IUP repository directory, or by creating a new repository. In either case, the directory is the same. In this case:

C:\Zynq_Book\hls_nco\solution1\impl\ip

We have now completed the generation of the NCO component as an IP Integrator compatible AXI-Lite block. You should now be familiar with:

- Specifying directives in Vivado HLS designs which define the control interface of the exported RTL.
- The process of specifying an AXI4 interface for a design, to enable a Vivado HLS system to be easily connected to the Zynq PS.
- Exporting a Vivado HLS design as an IP core that is compatible with the Vivado IP Catalog and IP Integrator.

The Zynq Book Tutorials 5

Adventures with IP Integrator

v1.4, June 2015

Revision History

Date	Version	Changes
22/10/2013	1.0	First release for Vivado Design Suite version 2013.3
28/01/2014	1.1	Updated for changes in Vivado Design Suite version 2013.4
06/05/2014	1.2	Updated for changes in Vivado Design Suite version 2014.1
10/09/2014	1.2.1	Minor corrections.
10/04/2015	1.3	Updated for changes in Vivado Design Suite version 2014.4
05/05/2015	1.3.1	Updated to include Zybo development board for Vivado Design Suite version 2014.4
19/06/2015	1.4	Updated for changes in Vivado Design Suite version 2015.1

Introduction

In this tutorial you will bring together all of the custom IP modules that you created in the previous set of practical exercises, along with other IP from the Vivado IP Catalog, to create a DSP system for implementation on a Zynq development board (note that all IP required by this design is also provided separately). IP for the control of the audio codec on the Zynq development board will be introduced and all modifications to the IP Integrator design will be carried out. A software application will be developed in the SDK which will configure all of the IP modules and control the interactions between them and the PS.

The tutorial is split into three exercises as follows:

Exercise 5A - This exercise focuses on importing all of the custom IP modules into the Vivado IP Catalog for inclusion in an IP Integrator DSP design. The individual IP blocks will be explored, along with their customisable parameters.

Exercise 5B - The Analog Devices ADAU1761 audio codec on the ZedBoard and SSM2603 audio codec on the Zybo will be introduced in this exercise, with the inclusion of some pre-packaged IP. Both IPs implement a I^2S serial communication for sending and receiving audio samples to/from the audio codec. The audio samples are transferred between the PL and the PS via a standard AXI-Lite connection. In order to use the audio codec, a variety of modifications must be made to the Zynq PS, such as the inclusion of second fabric clock to drive the codec, and the enabling of a I^2C interface for the communication of control signals between the PS and the codec.

In order to map the external interfaces in the design to physical pins on the Zynq device, a Xilinx Design Constraints (XDC) file must be created and included in the design. This informs the synthesis and implementation processes in Vivado where to route the external interface signals. The format of the XDC file will be explored before generating the hardware for the finalised design.

Exercise 5C - In this final exercise, the finalised design from Exercise 5B will be exported to the SDK for software development. Here, the application which will control the interactions between the various custom IP modules, the PS and the audio codec will be created. The various software driver files will also be explored before building and running the application on the Zynq development board for testing.

NOTE: Exercise 5C requires you to be able to send keyboard commands to the Zynq PS via the UART terminal. To do this, it is necessary to use third-party terminal program. In this tutorial, we shall be using PuTTY which can be downloaded for free from the following link:

http://www.chiark.greenend.org.uk/~sgtatham/putty/download.html

To download the standalone executable, select the **putty.exe** download from the **Binaries** section.

Exercise 5A Importing IP to the Vivado IP Catalog

In this exercise we will be concentrating on importing existing custom IP into the Vivado IP Catalog. We will be importing the various IP blocks that we created in **The Zynq Book Tutorial IP Creation**.

We will start by creating a new Vivado Project.

(a) Launch Vivado 2015.1 and create a new RTL project called ***adventures_with_ip*** in the **C:\Zynq_Book** directory, ensuring that the option to ***Create project subdirectory*** is selected. Select ***VHDL*** as the *Target language* and the appropriate part for your **Zynq** development board.

(b) From *Flow Navigator*, select ***IP Catalog*** from the **Project Manager** section.

The **IP Catalog** will open in the *Workspace*, as seen in Figure 5.1. Note the position of the **IP Settings** button which we will need shortly.

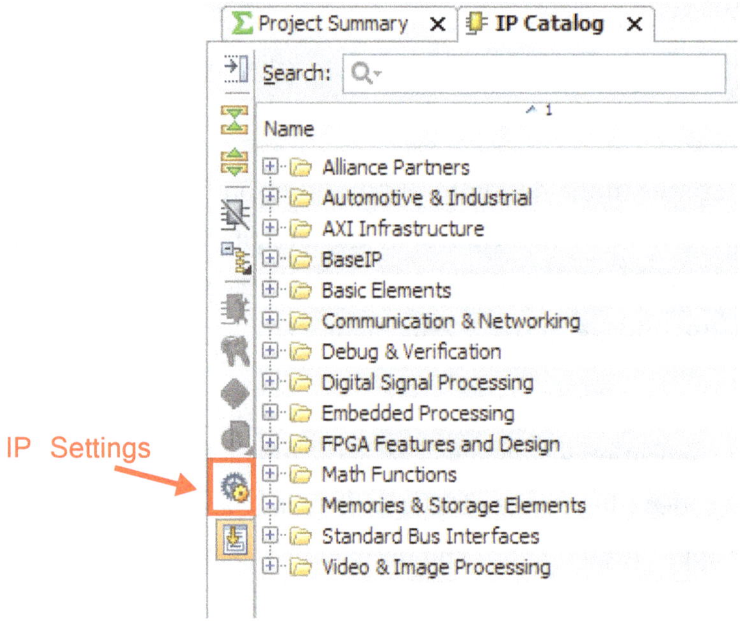

Figure 5.1: Vivado IP Catalog

In order to import our custom IP into the IP Catalog, we must add a new software repository to the IP Catalog. We will create a new directory to act as our IP repository and all of our IP sources to it.

(c) In Windows Explorer, navigate to the location: **C:\Zynq_Book\ip_repo**. This is the IP repository that we created in **Tutorial 4**.

We must now add each of the IP sources that we created in **The Zynq Book Tutorial IP Creation** to our repository.

As the LED controller IP is already present in the IP repository, we do not need to import it.

(d) Open a second Windows Explorer and navigate to **C:\Zynq_Book\hdl_coder_lms\hdl_prj\ipcore\lms_pcore_v1_00_a**. Copy the archived IP ZIP file, ***ac.uk_user_lms_pcore_1.0.zip*** to the **ip_repo** directory.

(e) In the second Windows Explorer, navigate to **C:\Zynq_Book\hls_nco\solution1\impl\ip** and copy the archived IP ZIP file, ***xilinx_com_hls_nco_1_0.zip*** to the **ip_repo** directory.

That completes the copying of our custom made IP sources to our newly created IP repository.

(f) We will now add one more IP source to our repository — an existing IP block which controls the audio codec on the Zynq development board.

Zed

In Windows Explorer, navigate to **C:\Zynq_Book\sources\zedboard\adventures_with_ip_integrator\ip** and copy the archived IP ZIP file, ***zed_audio_ctrl.zip*** to the **ip_repo** directory that we located in Step (c).

If you have not completed the previous tutorial, a master set of the IP sources is contained in **C:\Zynq_Book\sources\zedboard\adventures_with_ip_integrator\ip** which you can copy into the repository for use in this tutorial.

Zybo

In Windows Explorer, navigate to **C:\Zynq_Book\sources\zybo\adventures_with_ip_integrator\ip** and copy the archived IP ZIP file, ***xilinx_com_zybo_audio_ctrl_1.0.zip*** to the **ip_repo** directory that we located in Step (c).

If you have not completed the previous tutorial, a master set of the IP sources is contained in **C:\Zynq_Book\sources\zybo\adventures_with_ip_integrator\ip** which you can copy into the repository for use in this tutorial.

Resume Now that we have created the IP repository and added all of our existing IP sources, we can now add the repository to the IP Catalog.

(g) In the Vivado IP Catalog tab, click the ***IP Settings*** button, , as highlighted in Figure 5.1.

The IP Settings window will open, as shown in Figure 5.2.

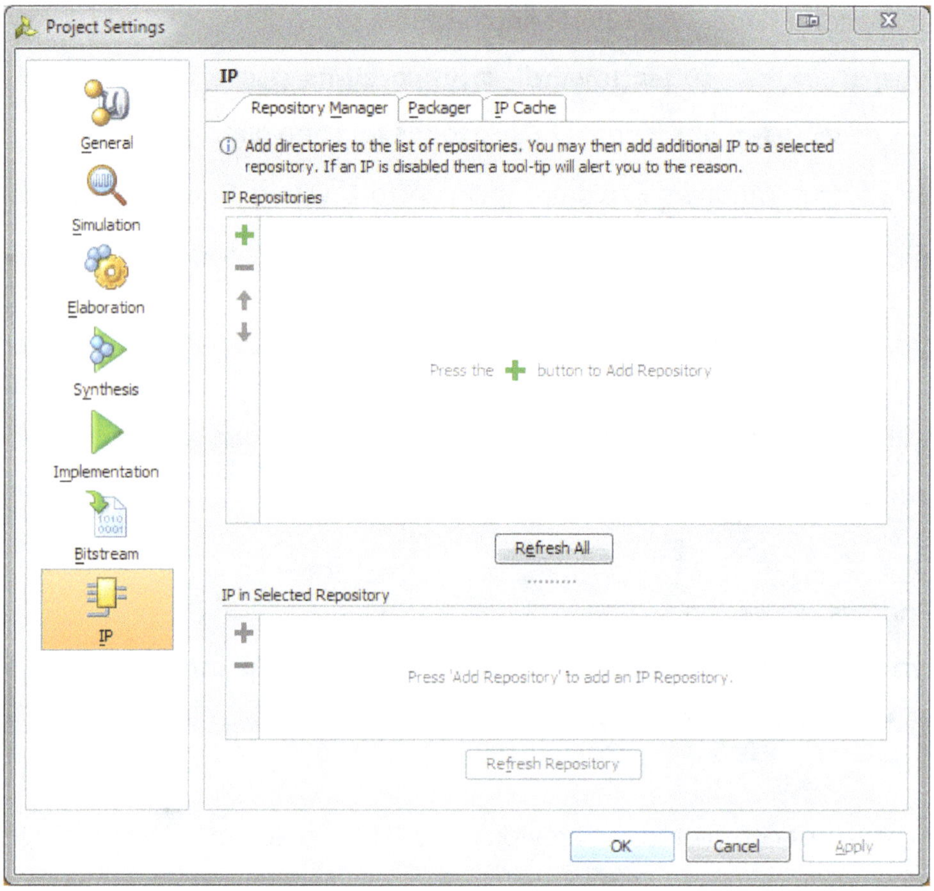

Figure 5.2: IP Settings Window

(h) Click the symbol in the *IP Repositories* panel, and browse to **C:\Zynq_Book\ip_repo**.

Click ***Select*** to add the repository to the IP Catalog.

You should see that the LED Controller IP is already present in the *IP in Selected Repository* pane as it is in un-archived format.

(i) We must now add the other IP sources to the repository by un-archiving them. In the Selected Repository panel, shown in Figure 5.2, click the + symbol to add IP.

Zed

The ***Select IP TO Add To Repository*** window will open as in Figure 5.3.

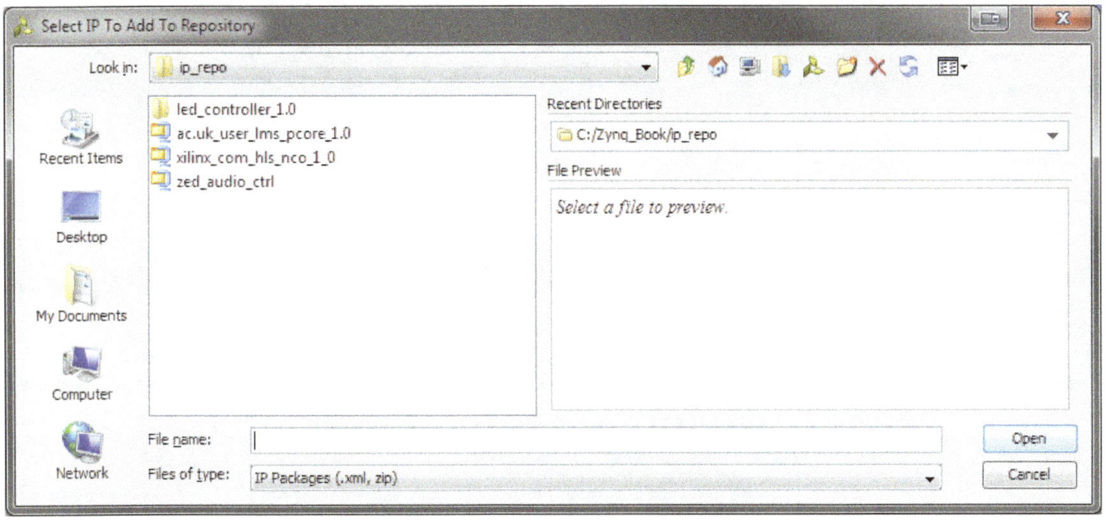

Figure 5.3: Select IP to Add to Repository

Select **ac.uk_user_lms_pcore_1.0.zip** and click **OK**. This will extract the archived IP sources into a usable format in the repository.

Repeat this procedure for the remaining IP sources:

- **xilinx_com_hls_nco_1_0.zip**
- **zed_audio_ctrl.zip**

The resulting IP in Selected Repository panel should resemble that shown in Figure 5.4.

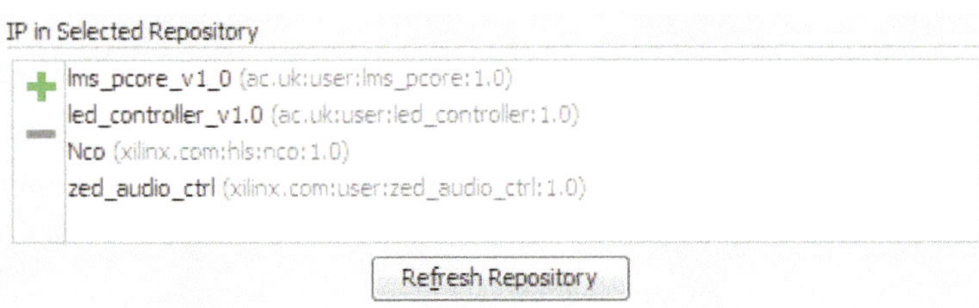

Figure 5.4: All IP sources added to IP Catalog

Click **OK**.

Zybo The ***Select IP TO Add To Repository*** window will open as in Figure 5.5.

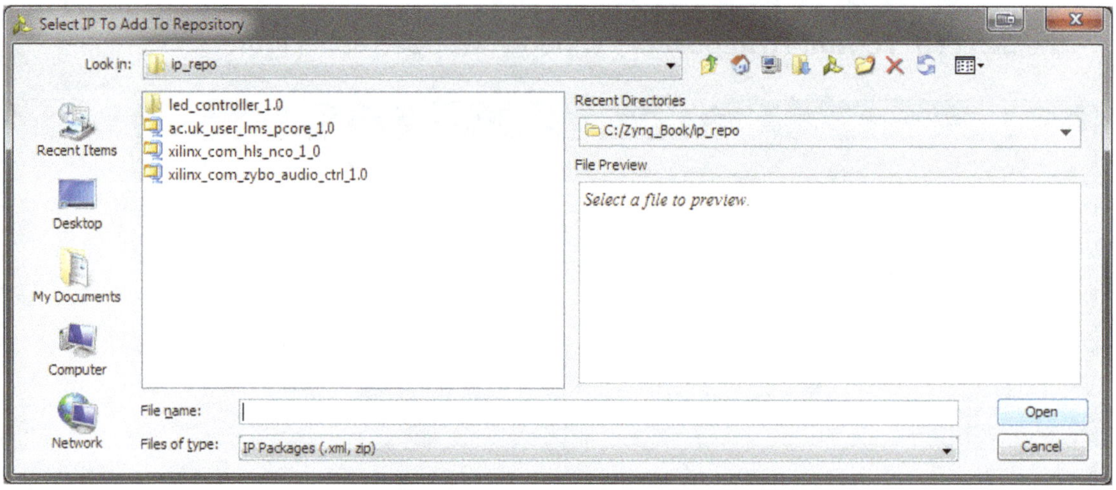

Figure 5.5: Select IP to Add to Repository

Select **ac.uk_user_lms_pcore_1.0.zip** and click **OK**. This will extract the archived IP sources into a usable format in the repository.

Repeat this procedure for the remaining IP sources:

- **xilinx_com_hls_nco_1_0.zip**
- **xilinx_com_zybo_audio_ctrl_1.0.zip**

The resulting IP in Selected Repository panel should resemble that shown in Figure 5.6.

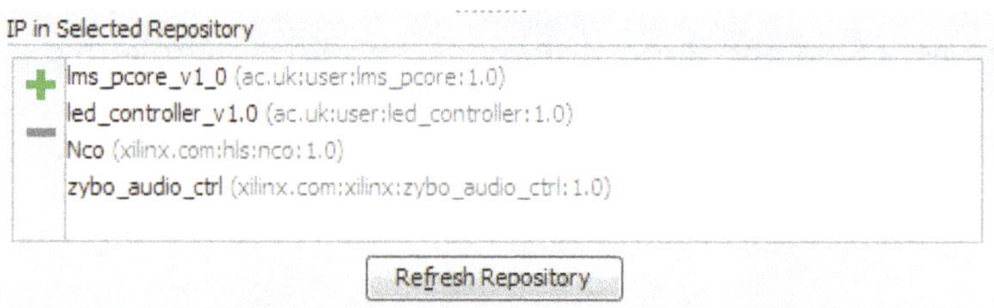

Figure 5.6: All IP sources added to IP Catalog

Click **OK**.

Resume With all of our IP now imported into the IP Catalog, we can now create an IP Integrator block design which incorporates all of the IP blocks.

(j) In *Flow Navigator*, select ***Create Block Design***.

(k) In the *Create Block Design* window, set the **Design name** as ***ip_design***, and click ***OK***.

(l) In the block design canvas, right-click and select ***Add IP***.

In the Search box, enter ***led_controller*** and double-click ***led_controller_v1_0*** to add an instance of the LED controller IP to the design.

(m) Repeat Step (l) searching for:

- ***nco*** and double-clicking ***Nco***
- ***lms*** and double-clicking ***lms_pcore_v1_0***

We have now added all of the custom IP that we created in the previous tutorial. At this point we will avoid adding the audio controller IP, as it is the focus of the next exercise.

In order to connect and control all of the IP, we must now add an instance of a Zynq Processor.

(n) In the block design canvas, right-click and select ***Add IP***.

In the Search box, enter ***zynq*** and double-click ***ZYNQ7 Processing System***.

At this stage, Designer Assistance should be available:

Zed

Select the ***Run Block Automation*** option from the *Designer Assistance* message at the top of the Diagram window. Select ***OK***, ensuring that the option to ***Apply Board Preset*** is selected, to generate the external connections for both the **DDR** and **FIXED_IO** interfaces, and apply the relevant board presets.

Your ZYNQ7 Processing System block should now resemble Figure 5.7.

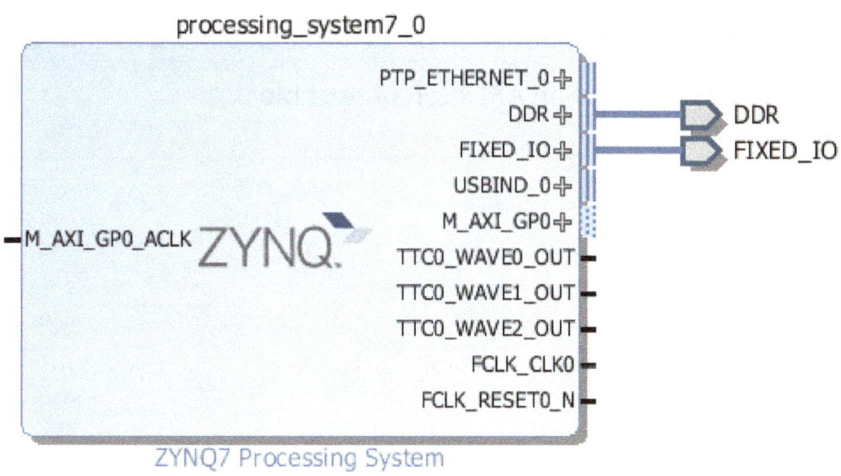

Figure 5.7: Zedboard ZYNQ7 Processing System External Connections

As the **Zedboard** platform is the target development board, and this was specified on creation of the project, Vivado will configure the Zynq processor block accordingly.

Zybo

Select the ***Run Block Automation*** option from the *Designer Assistance* message at the top of the Diagram window. Select ***OK***, ensuring that the option to ***Apply Board Preset*** is selected, to generate the external connections for both the **DDR** and **FIXED_IO** interfaces, and apply the relevant board presets.

Your ZYNQ7 Processing System block should now resemble Figure 5.8.

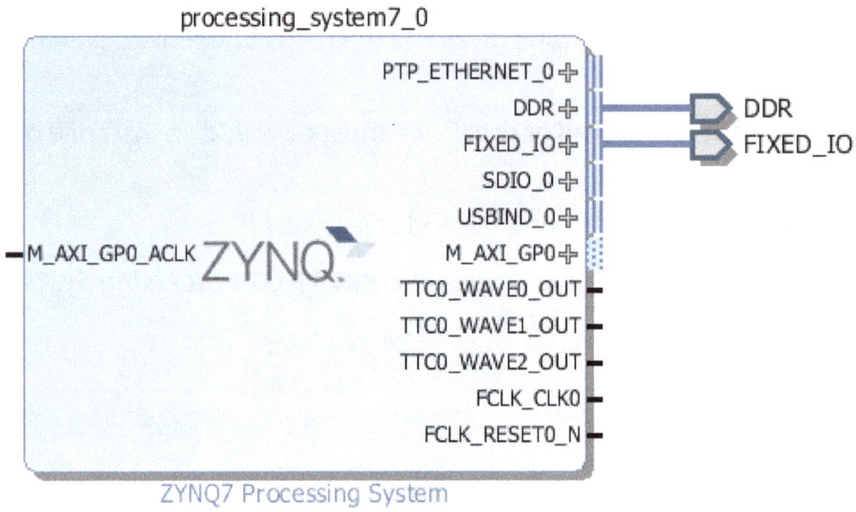

Figure 5.8: Zybo ZYNQ7 Processing System External Connections

As the **Zybo** platform is the target development board, and this was specified on creation of the project, Vivado will configure the Zynq processor block accordingly.

Resume

Run Connection Automation for each of the three IP blocks, to connect them to the Zynq7 Processing System block, via an AXI Interconnect block.

Select **All Automation** as in Figure 5.9 and click **OK**.

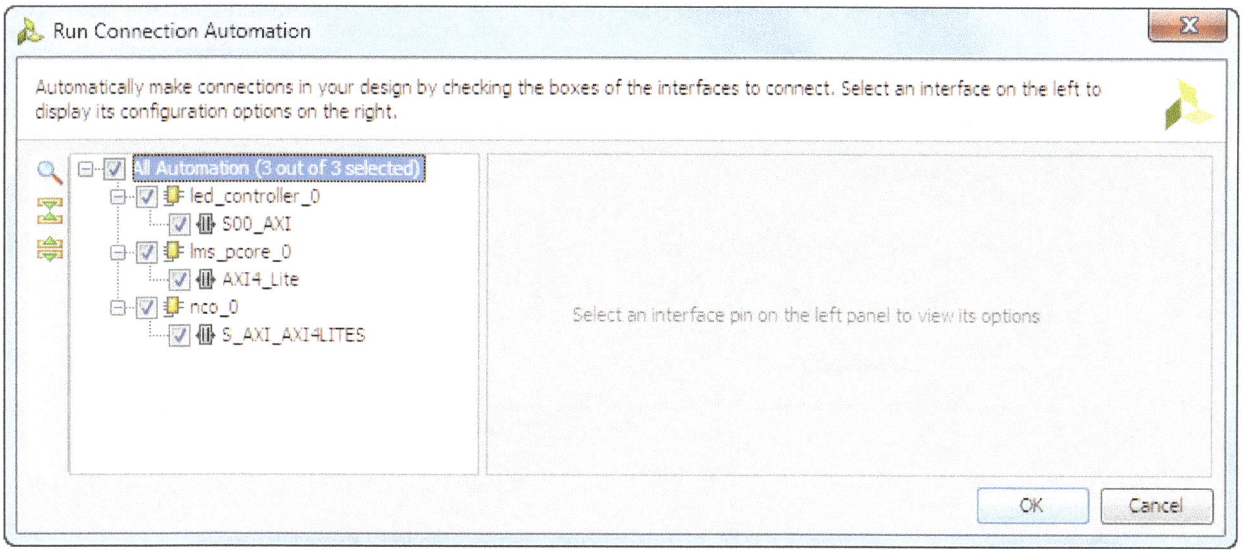

Figure 5.9: Run Connection Automation for three hardware blocks

You may recall that to allow the LED Controller block to control the LEDs on the board, the **LEDs_out** port must be made external.

(o) Hover the mouse pointer over the **LEDs_out** interface on the **led_controller** block until the cursor changes to a pencil. Right-click and select ***Make External***. Alternatively, select the interface and use the keyboard shortcut ***Ctrl+T***.

Notice that the **lms_pcore_0** block has two unconnected input ports, as highlighted in Figure 5.10.

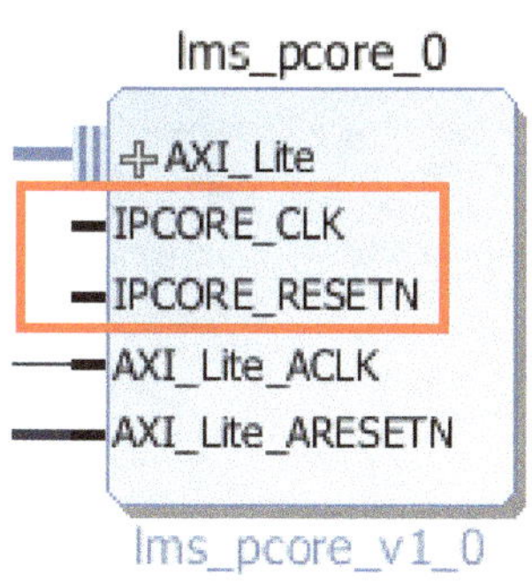

Figure 5.10: LMS IP block

These are the CLK and reset ports of the IP, and must be connected in order for the IP to be functional.

(p) Hover the mouse pointer over the **IPCORE_CLK** interface on the **lms_pcore_0** block until the cursor changes to a pencil. Click and drag the mouse pointer until it is hovering over the wire that connects to the **AXI_Lite_ACLK** interface and the wire is highlighted, as shown in Figure 5.11, and release the mouse button to create the connection.

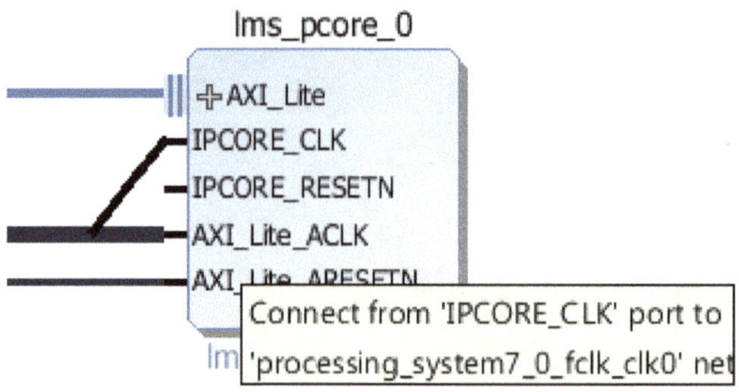

Figure 5.11: Manually connecting the LMS IP CLK

You should also see a pop-up message notifying you of the net which you are connecting to.

(q) Repeat the procedure of the previous step to, this time, connect the **IPCORE_RESETN** interface to the wire which connects to the **AXI_Lite_ARESETN** interface.

Zed Your current block diagram should now resemble Figure 5.12.

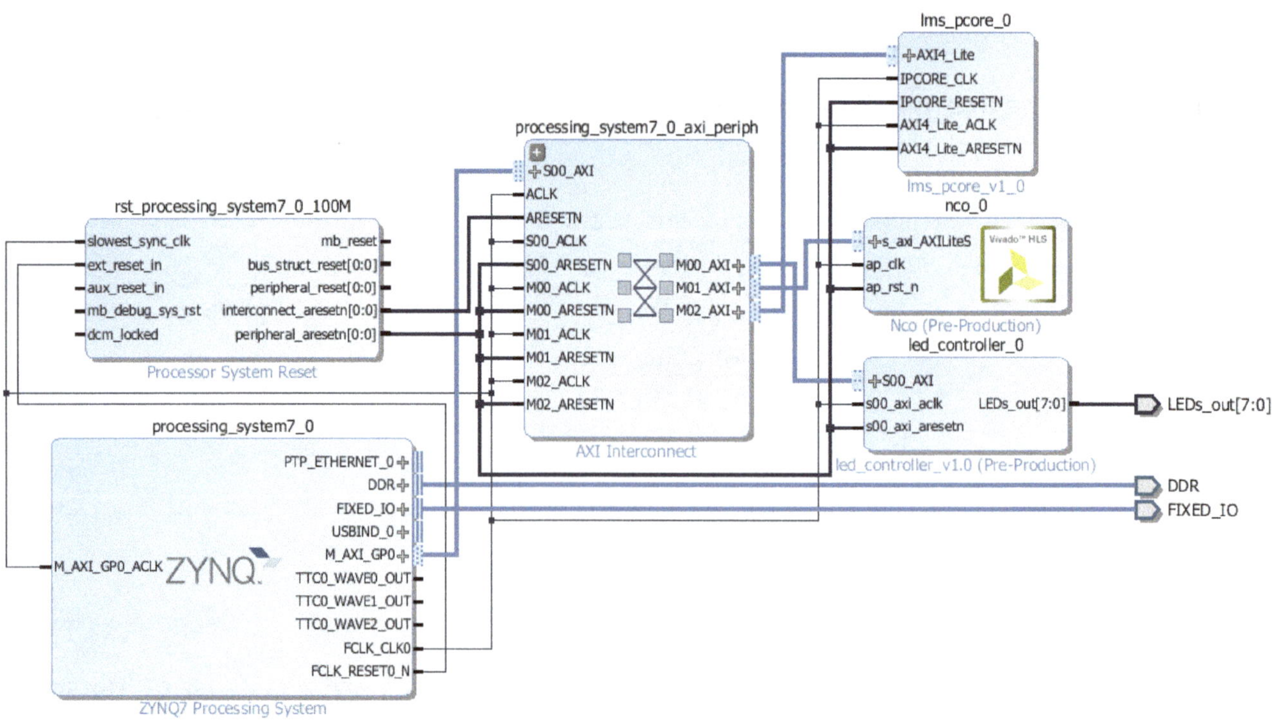

Figure 5.12: Zedboard end of exercise block diagram

Zybo Your current block diagram should now resemble Figure 5.13.

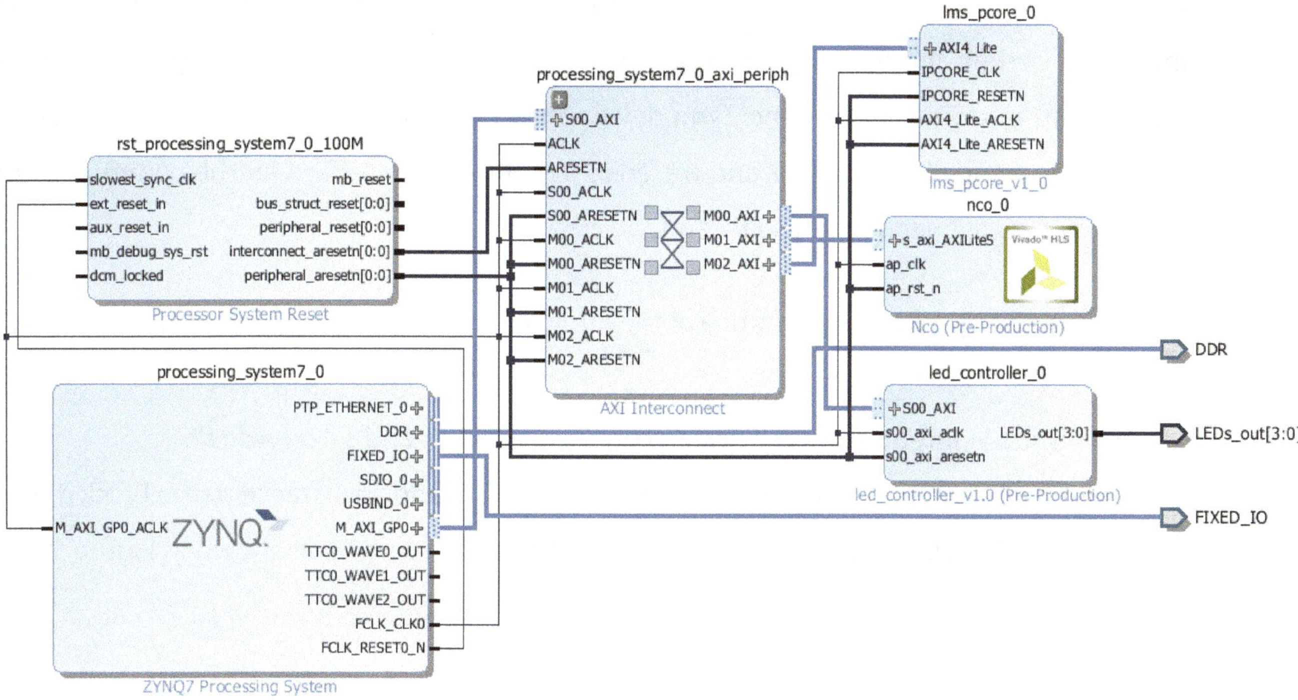

Figure 5.13: Zybo end of exercise block diagram

Resume At this stage we must now add and configure the audio controller IP, and so we will conclude this first exercise on importing custom IP to the Vivado IP Catalog. You should now be familiar with:

- Adding an IP repository to the Vivado IP Catalog.
- Importing and adding archived IP files to a custom IP repository.
- Adding custom IP to a Vivado IP Integrator block design.

Note: Do not close the current Vivado project as we will be using it again in the next exercise.

Exercise 5B Audio in Vivado IP Integrator

In this exercise we will be focusing on adding an audio controller IP instance to an existing Vivado IP Integrator design, and the modifications which must be made to the Zynq Processor block in order to use the audio codec on the Zynq development board. Such modifications include the addition of a second PL fabric clock and the enabling of the I^2C interface for the communication of control signals between the Zynq PS and the codec.

(a) We will begin by adding an instance of the audio controller IP to the block design.

Zed

In the Vivado IP Integrator block design canvas, right-click and select ***Add IP***.
Search for **audio** and double-click on ***zed_audio_ctrl***, to add an instance to the block design.
The ***zed_audio_ctrl_0*** block should now be visible on the canvas, as shown in Figure 5.14.

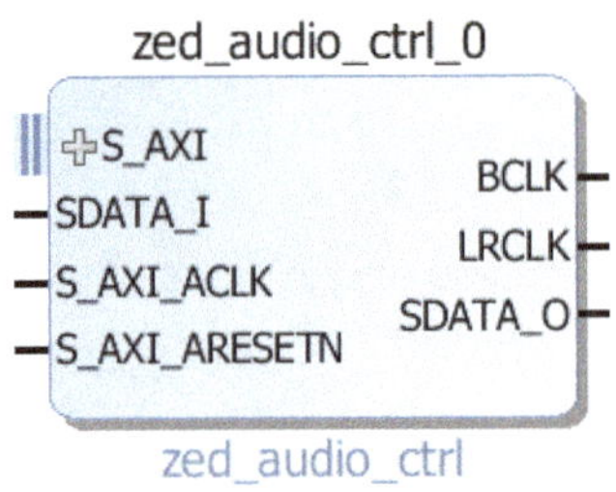

Figure 5.14: ZedBoard Audio Controller block

Make the initial connection between the *Zynq PS* and the *zed_audio_ctrl_0* block by clicking ***Run Connection Automation*** and clicking **OK**.

You should notice that there are still four unconnected ports. These are required to be made external to connect to the physical pins of the ZedBoard's audio codec.

Hover the mouse pointer over each of the unconnected interfaces on the **zed_audio_ctrl** block until the cursor changes to a pencil. Right-click and select ***Make External***. Alternatively, select the interface and use the keyboard shortcut ***Ctrl+T***.

Zybo

In the Vivado IP Integrator block design canvas, right-click and select ***Add IP***.
Search for **audio** and double-click on ***zybo_audio_ctrl***, to add an instance to the block design.

The ***zybo_audio_ctrl_0*** block should now be visible on the canvas, as shown in Figure 5.15.

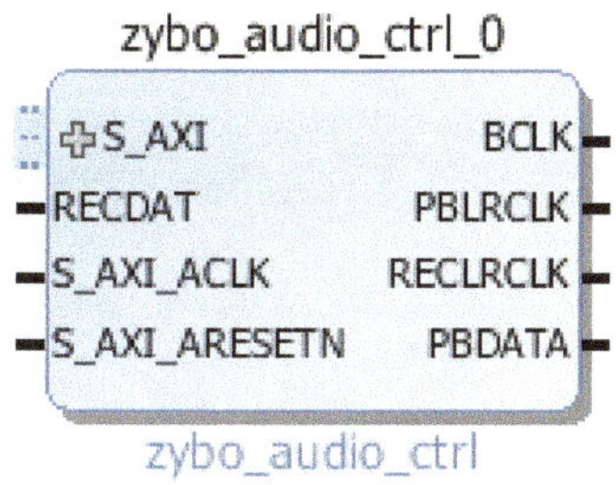

Figure 5.15: Zybo Audio Controller block

Make the initial connection between the *Zynq PS* and the *zybo_audio_ctrl_0* block by clicking ***Run Connection Automation*** and clicking **OK**.

You should notice that there are still five unconnected ports. These are required to be made external to connect to the physical pins of the Zybo's audio codec.

Hover the mouse pointer over each of the unconnected interfaces on the **zybo_audio_ctrl** block until the cursor changes to a pencil. Right-click and select ***Make External***. Alternatively, select the interface and use the keyboard shortcut ***Ctrl+T***.

Resume

The next step is to make the necessary modifications to the Zynq7 PS block.

(b) Double-click on the ***Zynq7 Processing System*** block to open the Re-customize IP window, as shown in Figure 5.16.

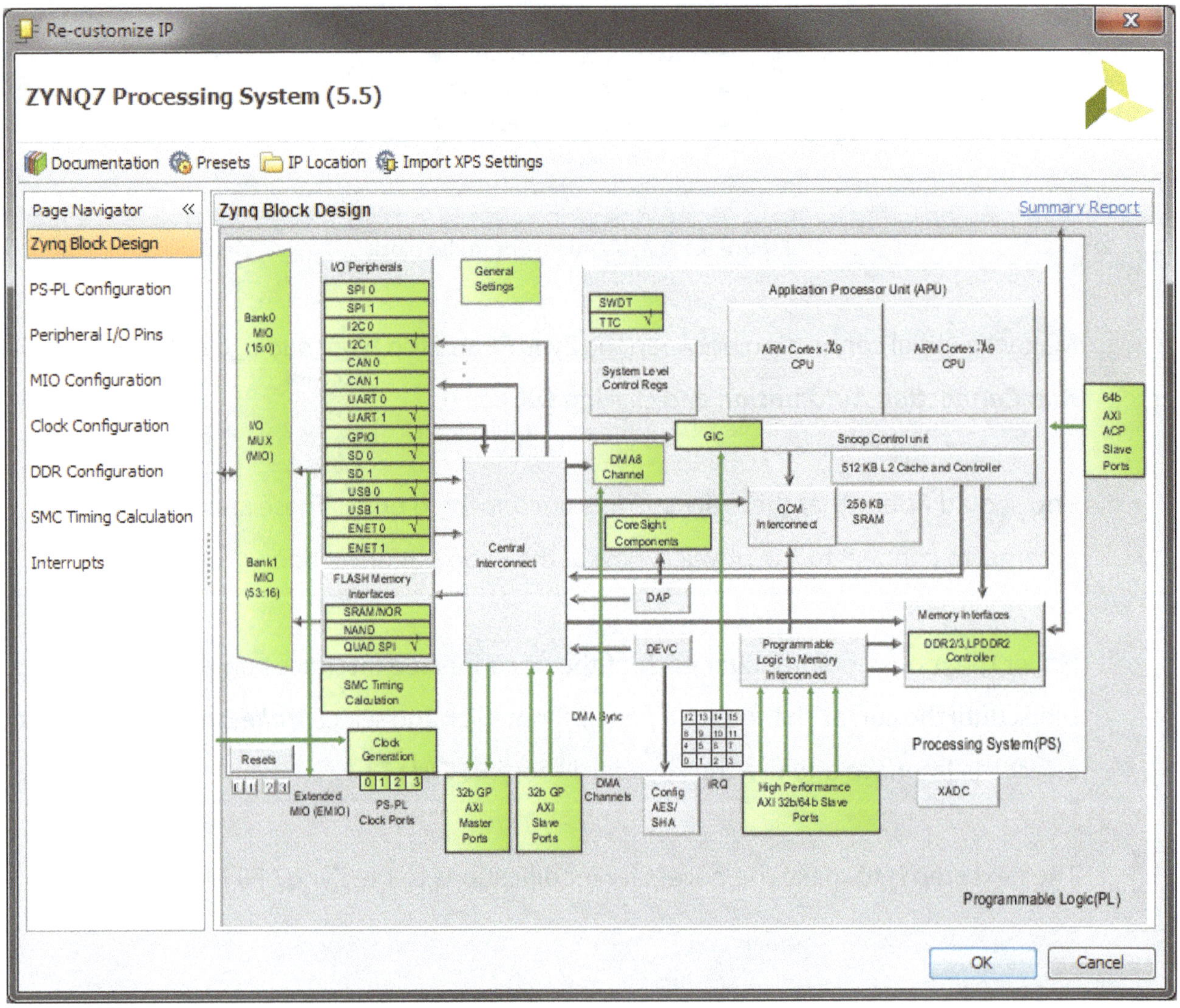

Figure 5.16: Re-customize IP window for Zynq PS

This view allows you to make changes to the configuration of the Zynq PS. As IP Integrator is *board aware*, all of the basic settings that apply to many Zynq development boards have been made for us. There are a few changes, however, that must be made when using the audio codec.

First we will add a second PL fabric clock as a separate clock is required for the **MCLK** pin on the audio codec.

(c) Click on ***Clock Configuration*** in the *Page Navigator* panel on the left hand side of the window. Expand ***PL Fabric*** clocks in the *Clock Configuration* panel, and enable ***FCLK_CLK1***.

Zed Change the ***Requested Frequency*** of **FCLK_CLK1** to ***10 MHz***, as shown in Figure 5.17.

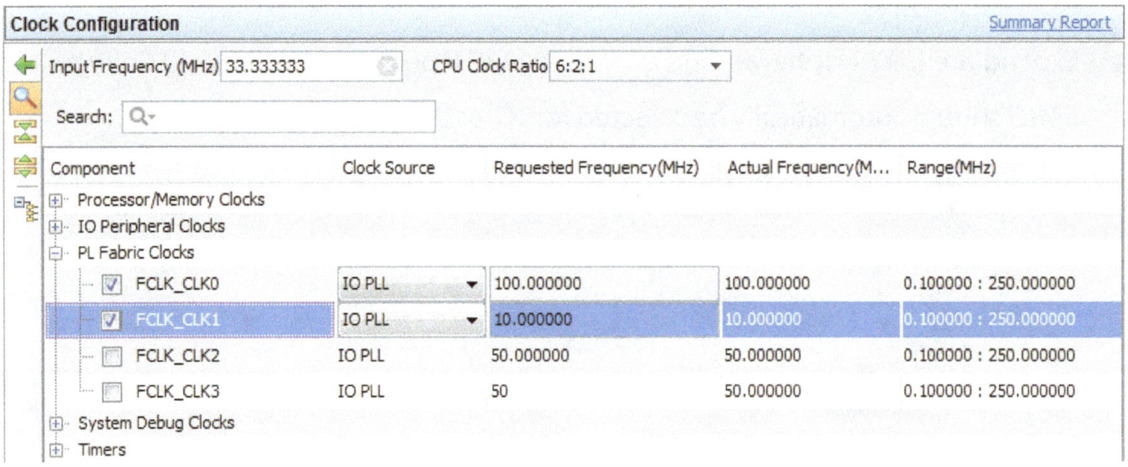

Figure 5.17: Adding a 10 MHz fabric clock

Zybo Change the ***Requested Frequency*** of **FCLK_CLK1** to ***12.288 MHz***, as shown in Figure 5.18.

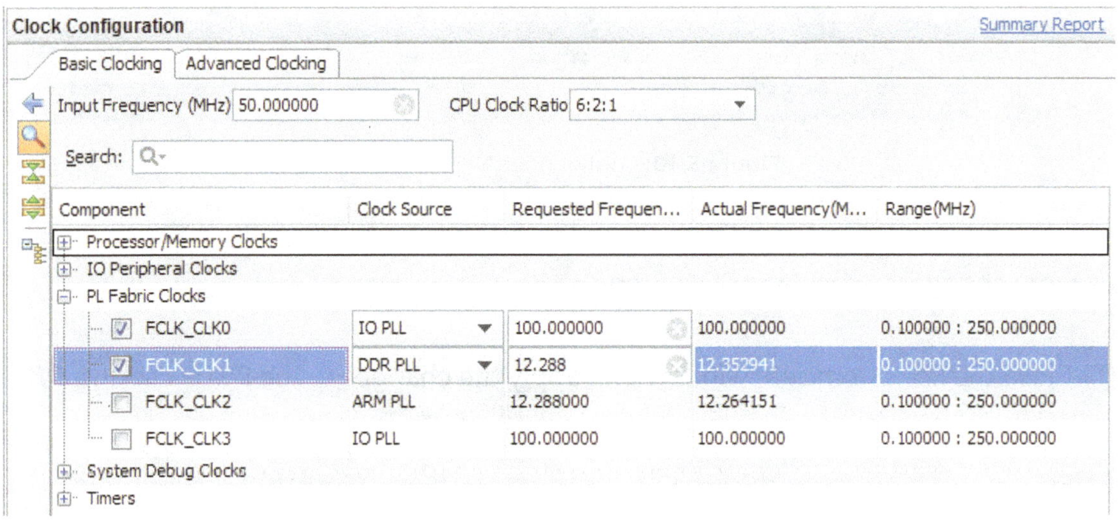

Figure 5.18: Adding a 12.288 MHz fabric clock

Resume Next, we must enable one of the Zynq PS's I^{2}C communication interfaces to allow the PS to communicate with the audio codec.

(d) Select ***MIO Configuration*** from the *Page Navigator* panel.

This configuration view allows us to enable/disable the PS peripherals. These peripherals can be routed through the dedicated **Multiplexed I/Os (MIO)** on the device, or through the **Extended Multiplexed I/Os (EMIOs)** which route to the PL fabric.

As we want to communicate with the audio codec (which is connected to fabric pins of the Zynq device) we will be routing the I^2C signals through the EMIOs.

(e) Expand the **I/O Peripherals** and enable the ***I2C 0*** peripheral in the *MIO Configuration* panel. ***EMIO*** should automatically be selected for **IO**, as shown in Figure 5.19.

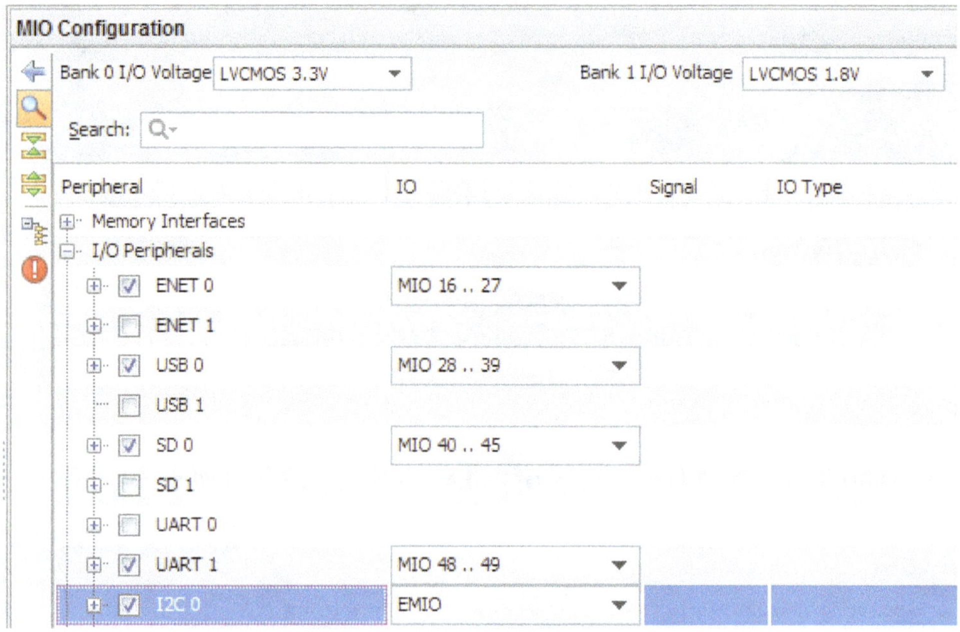

Figure 5.19: Configuring the I2C interface

No more changes to the Zynq PS are required.

(f) Close the *Re-customize IP* window and apply the changes to the PS by clicking ***OK***.

Zed

The IP Integrator canvas should update, and the ZYNQ7 Processing System block should now look like Figure 5.20.

You should note the addition of the two new interfaces, **IIC_0** and **FCLK_CLK1**. As these will be driving signals on the audio codec, which is situated on the board (external to the Zynq device), we must make these external.

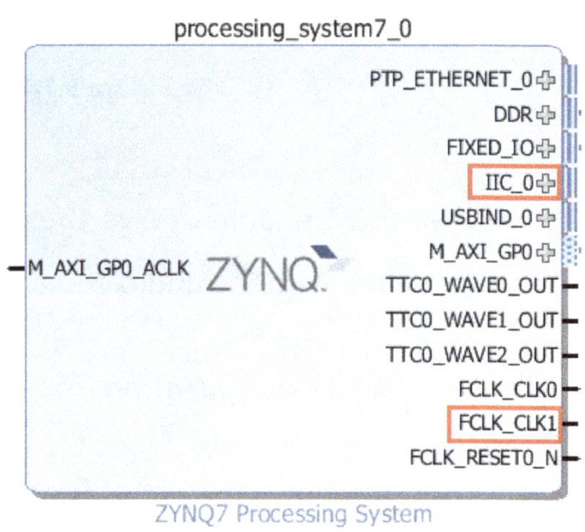

Figure 5.20: Zedboard Zynq7 Processing System block

(g) Hover the mouse pointer over each of the **IIC_0** and **FCLK_CLK1** interfaces on the processing_system7_0 block until the cursor changes to a pencil. Right-click and select ***Make External***. Alternatively, select the interface and use the keyboard shortcut ***Ctrl+T***.

Zybo

The IP Integrator canvas should update, and the ZYNQ7 Processing System block should now look like Figure 5.21.

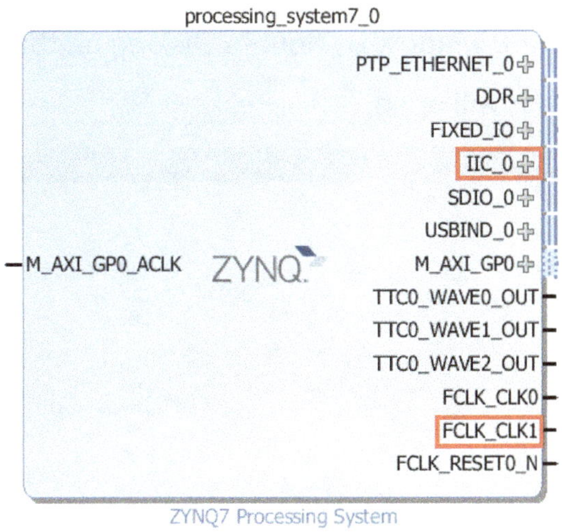

Figure 5.21: Zybo Zynq7 Processing System block

You should note the addition of the two new interfaces, **IIC_0** and **FCLK_CLK1**. As these will be driving signals on the audio codec, which is situated on the board (external to the Zynq device), we must make these external.

(h) Hover the mouse pointer over each of the **IIC_0** and **FCLK_CLK1** interfaces on the processing_system7_0 block until the cursor changes to a pencil. Right-click and select ***Make External***. Alternatively, select the interface and use the keyboard shortcut ***Ctrl+T***.

Resume

The final addition to the Block design that we need to make, is to add a single GPIO instance and a dual GPIO instance:

- Single-channel GPIO with a width of 2-bits to connect to the **Zedboard's** audio codec's *I2C ADDR* pins or a width of 1-bit to connect to the **Zybo's** audio codec's *Digital Mute*.
- Dual-channel GPIO with a width of 32-bits to connect to the push buttons and slide switches for user input.

First we will add the single GPIO to control the Zynq development board codec.

(i) In the Vivado IP Integrator block design canvas, right-click and select ***Add IP***.

Search for **gpio** and double-click on ***AXI_GPIO***, to add an instance to the block design.

(j) ***Run Connection Automation*** for the **axi_gpio_0/S_AXI** interface, to connect the GPIO controller to the Zynq PS via the ***AXI Interconnect*** (do not Run Connection Automation for the GPIO's output interface).

(k) Open the *Re-customize IP* window by double-clicking on the **axi_gpio_0** block. The window, as shown in Figure 5.22, will open.

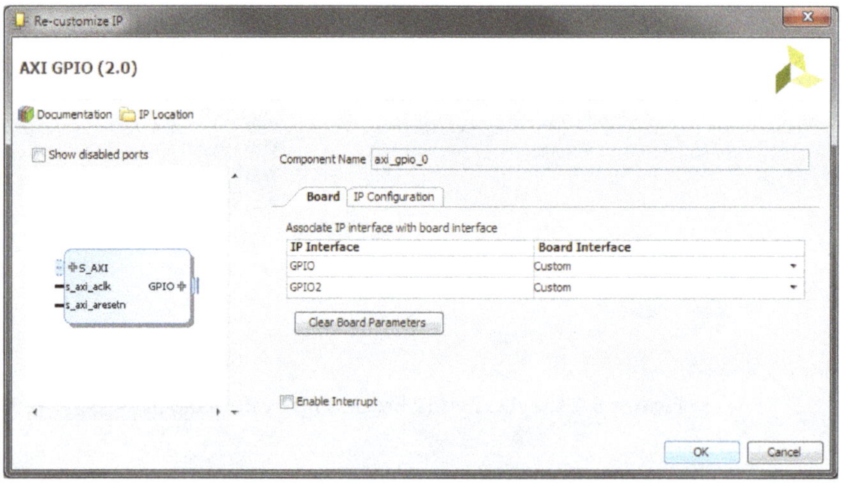

Figure 5.22: Re-customize IP window (GPIO)

(l) Select the ***IP Configuration*** tab.

Zed

Enter **2** as the **GPIO Width**, as shown in Figure 5.23, and close the window by clicking ***OK***.

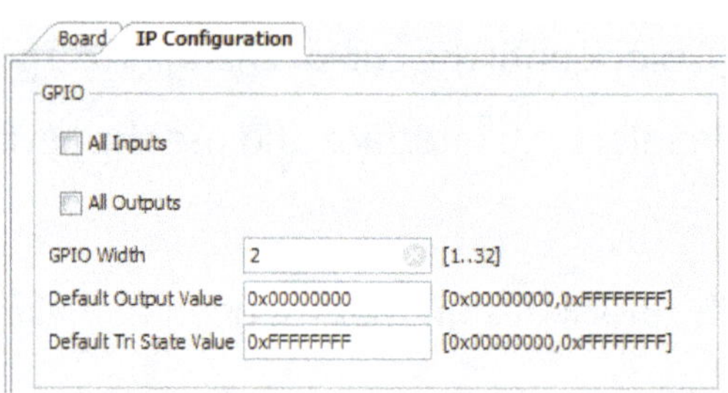

Figure 5.23: Zedboard GPIO width setting

Zybo

Enter ***1*** as the **GPIO Width**, as shown in Figure 5.24, and close the window by clicking ***OK***.

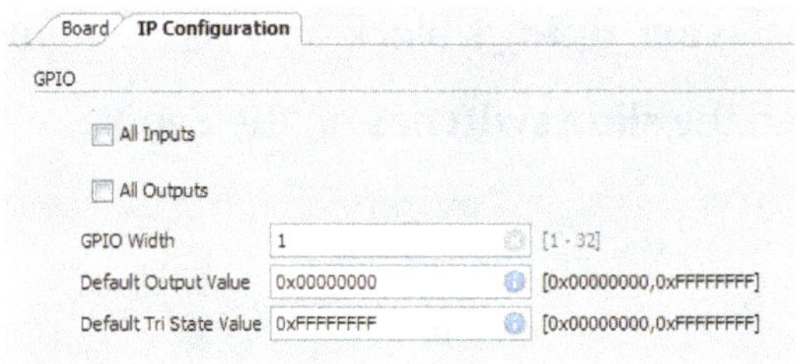

Figure 5.24: Zybo GPIO width setting

Resume (m) Make the ***GPIO*** interface of the **axi_gpio_0** block external.

Next we will add a second instance of the AXI GPIO Controller

(n) Add an instance of the ***AXI_GPIO*** IP to the block design and ***Run Connection Automation*** for **S_AXI** to connect the new GPIO controller to the Zynq PS via the ***AXI Interconnect*** (do not Run Connection Automation for the GPIO's output interface).

(o) The newly created **AXI GPIO** block must now be configured to allow for ***Dual Channel*** operation.

Double-click on the **axi_gpio_1** block to open the *Re-customize IP* window.

In the *IP Configuration* tab, select the option to ***Enable Dual Channel***, and click ***OK***.

Zed

You should see that the **axi_gpio_1** block now has two output ports, 1 each to connect to the **push buttons** and the **slide switches** on the ZedBoard:

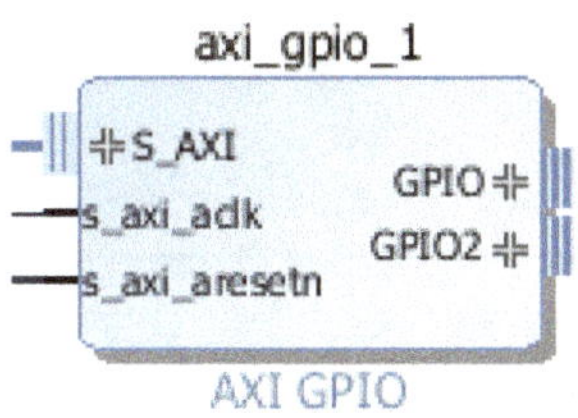

Run Connection Automation for **/axi_gpio_1/GPIO** and select ***btns_5bits*** as the option for **Select Board Interface**.

Click ***OK***.

Run Connection Automation for **/axi_gpio_1/GPIO2** and select ***sws_8bits*** as the option for **Select Board Interface**.

Click ***OK***.

Zybo

You should see that the **axi_gpio_1** block now has two output ports, 1 each to connect to the **push buttons** and the **slide switches** on the Zybo:

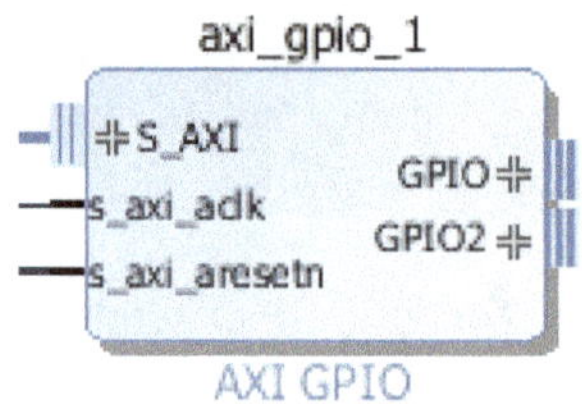

Run Connection Automation for **/axi_gpio_1/GPIO** and select ***btns_4bits*** as the option for **Select Board Interface**.

Click ***OK***.

Run Connection Automation for **/axi_gpio_1/GPIO2** and select ***sws_4bits*** as the option for **Select Board Interface**.

Click ***OK***.

Resume (p) The Zynq Processing System address of each IPCore will now be re-configured to increase their efficiency and reduce unused address space.

Zed

Select the ***Address Editor*** tab from the *Block Design* window, as highlighted in Figure 5.25.

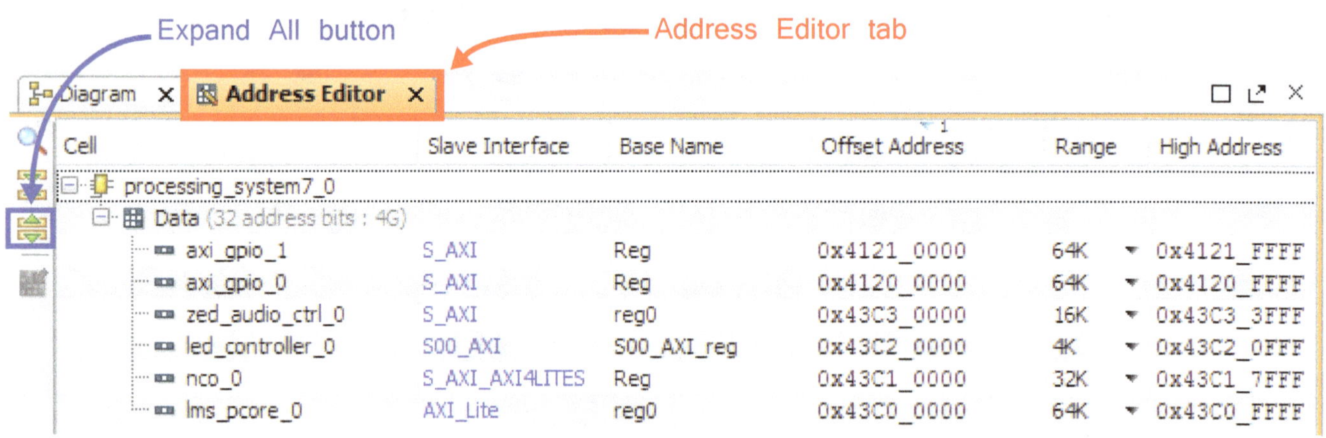

Figure 5.25: Zedboard Address Editor tab

Click the ***Expand All*** button, as highlighted in Figure 5.25.

Check the assigned *Range* for each of the peripheral cells against Figure 5.25.

If they do not match those in Figure 5.25, you must manually change the ranges so that they do match Figure 5.25. If they match those in Figure 5.25, you can skip this step and move on to **Step (q)**.

Zybo

Select the ***Address Editor*** tab from the *Block Design* window, as highlighted in Figure 5.26.

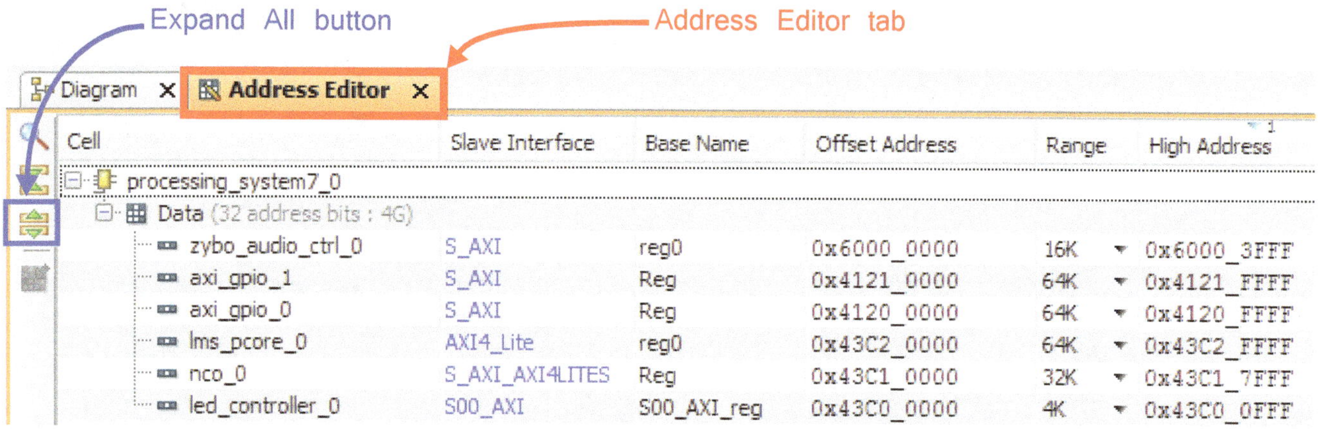

Figure 5.26: Zybo Address Editor tab

Click the ***Expand All*** button, as highlighted in Figure 5.26.

Check the assigned *Range* for each of the peripheral cells against Figure 5.26.

If they do not match those in Figure 5.26, you must manually change the ranges so that they do match Figure 5.26. If they match those in Figure 5.26, you can skip this step and move on to **Step (q)**.

Resume

(q) Return to the block design by selecting the ***Diagram*** tab in the *IP Integrator* window.

(r) Click the Regenerate Layout button, , to regenerate the layout of the various IP blocks and make the block design easier to follow.

Zed Your complete block design should be similar to Figure 5.27.

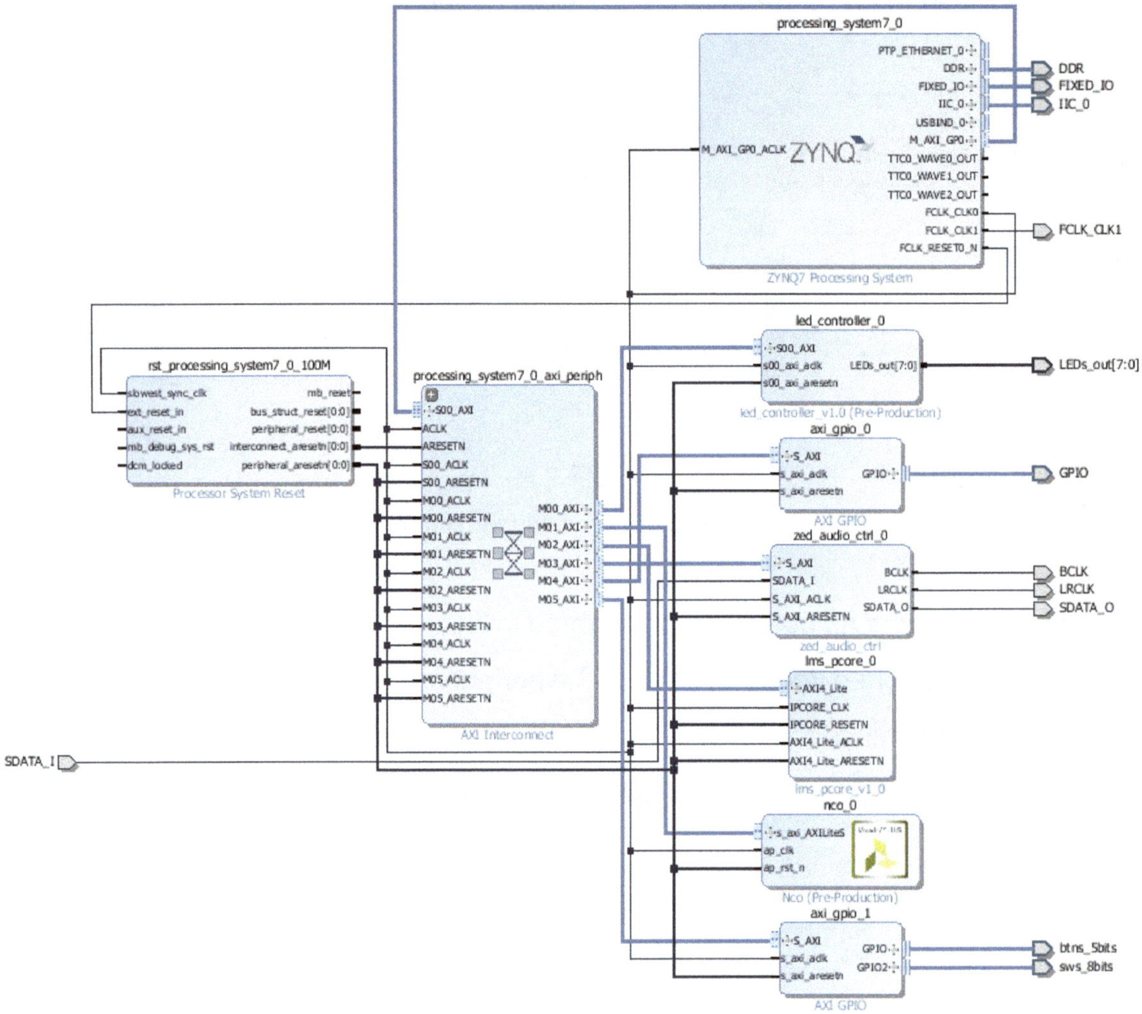

Figure 5.27: Zedboard Completed block design

Zybo

Your complete block design should be similar to Figure 5.28.

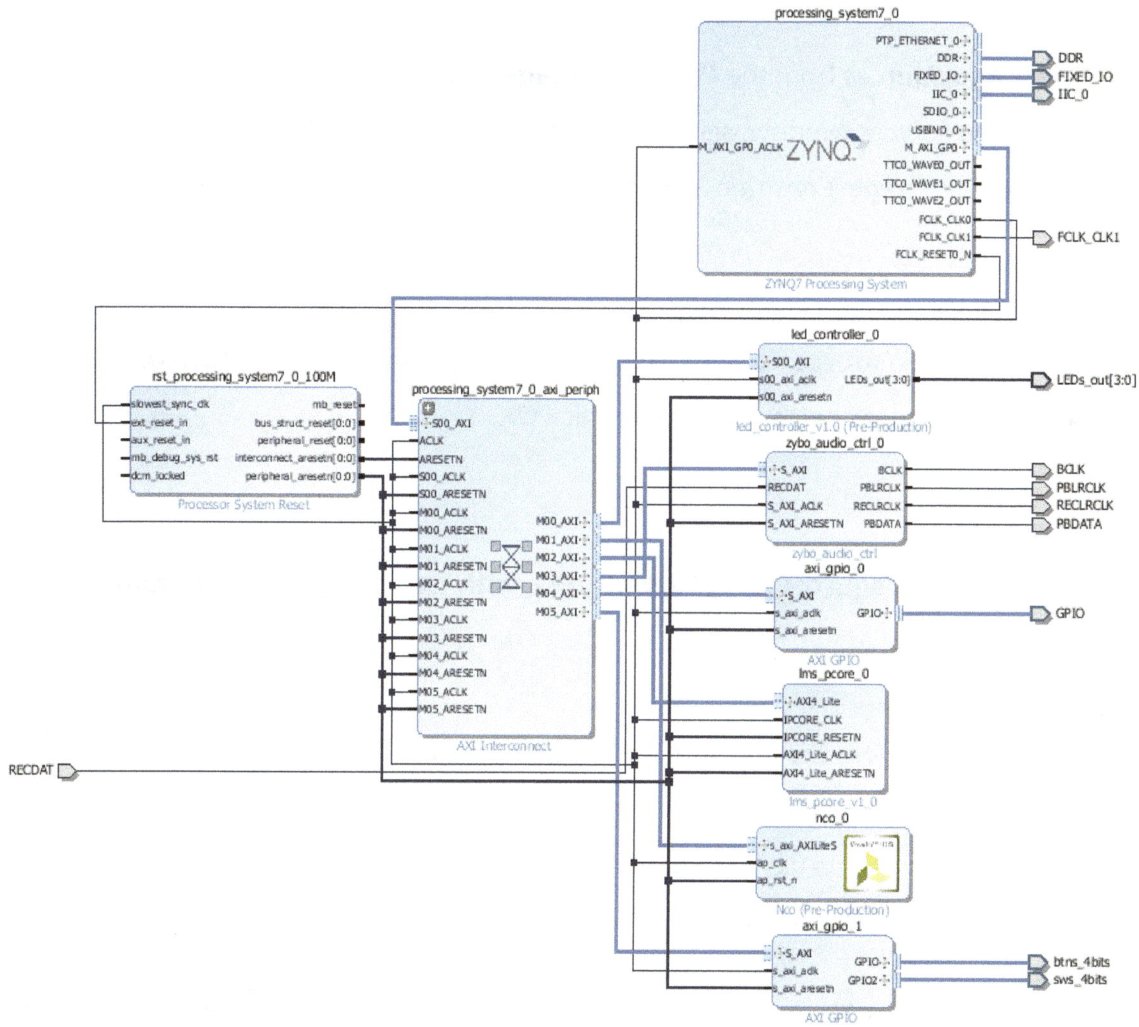

Figure 5.28: Zybo Completed block design

Resume (s) Save the block design.

Before we can run synthesis and implementation for our design, we must generate the RTL files for our block design.

(t) Generate a top-level HDL wrapper file, by right-clicking on **ip design** in the **Sources** tab and selecting ***Create HDL Wrapper***.

In the *Create HDL Wrapper* window, select **Let Vivado manage wrapper and auto-update**, and click ***OK***.

The next task that we have to do in Vivado before we can run synthesis and implementation of the design, is to add a constraints file which will map the external interfaces of our design to specific pins on the Zynq device.

(u) Select ***Add Sources*** from the **Project Manager** section of *Flow Navigator*.

In the *Add Sources* window, select ***Add or Create Constraints***, and click ***Next***.

In the *Add or Create Constraints* window, click the + symbol and then select ***Add Files...***

Zed

Navigate to:

C:/Zynq_Book/sources/zedboard/adventures_with_ip_integrator/constraints

Select **adventures_with_ip.xdc**, and click ***OK***.

Zybo

Navigate to:

C:/Zynq_Book/sources/zybo/adventures_with_ip_integrator/constraints

Select **adventures_with_ip.xdc**, and click ***OK***.

Resume

Click ***Finish*** to close the *Add Sources* window, and import the constraints file.

(v) Open the constraints file by expanding the *Constraints* section of **Sources** tab, and double-clicking on ***adventures_with_ip.xdc***.

The top section of the file contains the constraints which map the individual bits of the **LEDs_out** interface to the corresponding pins on the Zynq device, and you will have seen these before in the first exercise of the previous tutorial.

Zed The bottom section of the file, as shown in Figure 5.29, contains the constraints which map the various external ports of the design which relate to the audio codec, to their corresponding pins on the Zynq device.

```
# ZedBoard Audio Codec Constraints
set_property PACKAGE_PIN AA6 [get_ports BCLK]
set_property IOSTANDARD LVCMOS33 [get_ports BCLK]

set_property PACKAGE_PIN Y6 [get_ports LRCLK]
set_property IOSTANDARD LVCMOS33 [get_ports LRCLK]

set_property PACKAGE_PIN AA7 [get_ports SDATA_I]
set_property IOSTANDARD LVCMOS33 [get_ports SDATA_I]

set_property PACKAGE_PIN Y8 [get_ports SDATA_O]
set_property IOSTANDARD LVCMOS33 [get_ports SDATA_O]

#MCLK
set_property PACKAGE_PIN AB2 [get_ports FCLK_CLK1]
set_property IOSTANDARD LVCMOS33 [get_ports FCLK_CLK1]

set_property PACKAGE_PIN AB4 [get_ports iic_0_scl_io]
set_property IOSTANDARD LVCMOS33 [get_ports IIC_0_scl_io]

set_property PACKAGE_PIN AB5 [get_ports iic_0_sda_io]
set_property IOSTANDARD LVCMOS33 [get_ports IIC_0_sda_io]

set_property PACKAGE_PIN AB1 [get_ports {gpio_tri_io[0]}]
set_property IOSTANDARD LVCMOS33 [get_ports {GPIO_tri_io[0]}]

set_property PACKAGE_PIN Y5 [get_ports {gpio_tri_io[1]}]
set_property IOSTANDARD LVCMOS33 [get_ports {GPIO_tri_io[1]}]
```

Figure 5.29: ZedBoard audio codec constraints

Zybo The bottom section of the file, as shown in Figure 5.30, contains the constraints which map the various external ports of the design which relate to the audio codec, to their corresponding pins on the Zynq device.

```
## Zybo Audio Codec Constraints
set_property PACKAGE_PIN K18 [get_ports BCLK]
set_property IOSTANDARD LVCMOS33 [get_ports BCLK]

set_property PACKAGE_PIN L17 [get_ports PBLRCLK]
set_property IOSTANDARD LVCMOS33 [get_ports PBLRCLK]

set_property PACKAGE_PIN M18 [get_ports RECLRCLK]
set_property IOSTANDARD LVCMOS33 [get_ports RECLRCLK]

set_property PACKAGE_PIN K17 [get_ports RECDAT]
set_property IOSTANDARD LVCMOS33 [get_ports RECDAT]

set_property PACKAGE_PIN M17 [get_ports PBDATA]
set_property IOSTANDARD LVCMOS33 [get_ports PBDATA]

# MCLK
set_property PACKAGE_PIN T19 [get_ports FCLK_CLK1]
set_property IOSTANDARD LVCMOS33 [get_ports FCLK_CLK1]

#I2C 0 Interface
set_property PACKAGE_PIN N18 [get_ports iic_0_scl_io]
set_property IOSTANDARD LVCMOS33 [get_ports iic_0_scl_io]

set_property PACKAGE_PIN N17 [get_ports iic_0_sda_io]
set_property IOSTANDARD LVCMOS33 [get_ports iic_0_sda_io]

#GPIO_0[0] Digital Mute
set_property PACKAGE_PIN P18 [get_ports {gpio_tri_io[0]}]
set_property IOSTANDARD LVCMOS33 [get_ports {gpio_tri_io[0]}]
```

Figure 5.30: Zybo audio codec constraints

Resume Next, we will create a bitstream so that we can program the PL of the Zynq device with our design.

(w) In *Flow Navigator*, select ***Generate Bitstream*** from the **Program and Debug** section.

At the *No Implementation Results Available* window, click ***Yes*** to launch synthesis and implementation. This may take a few minutes depending on the speed of your computer.

When bitstream generation is complete, select ***Open Implemented Design*** in the *Bitstream Generation Completed* window, and click ***OK***.

Finally, we can export the hardware to the SDK, where we will create a software application to control the system in the next exercise.

(x) Select **File > Export > Export Hardware...** from the *Menu Bar*.

Ensure that the option to ***Include Bitstream*** is selected, and click ***OK***.

(y) Launch the SDK in Vivado by selecting ***File > Launch SDK*** from the *Menu Bar* and Click ***OK***.

This concludes this exercise of audio on a Zynq development board. You should now be familiar with:

- Making the required changes to the Zynq PS in order to use the audio codec on the ZedBoard and/or Zybo.
- Making the required external connections to allow the Zynq PL to be connected to the audio codec via the external Zynq device pins.
- Using a constraints file to map the external interfaces of the design which relate to the audio codec, to the corresponding pins on the Zynq device.

Exercise 5C Creating an Audio Software Application in SDK

In this final exercise we will be creating a software application which ties together all of the IP modules which we have created, to create a DSP-oriented system. The procedure of setting up the ZedBoard and Zybo audio codec via the hardware registers will also be introduced.

Once the SDK has launched from the previous exercise, we can start by creating a new application.

(a) Select **File > New > Application Project** from the *Menu Bar*.
    In the *New Project* dialogue, enter ***adventures_with_ip*** as the *Project name*.
    By default the option to create a new **Board Support Package** will be selected.
    Click ***Next***.

(b) In the *Templates* dialogue, select ***Empty Application***, and click ***Finish***.

You should recall that when we created the custom IP peripherals in the previous tutorial a set of software driver files were generated for each. We must now point SDK to those driver files. This is done by adding new repositories to the SDK project.

(c) Navigate to **Xilinx Tools > Repositories** in the *Menu Bar.*

In the *Repositories Preferences* window, click on ***New***, as shown in Figure 5.31.

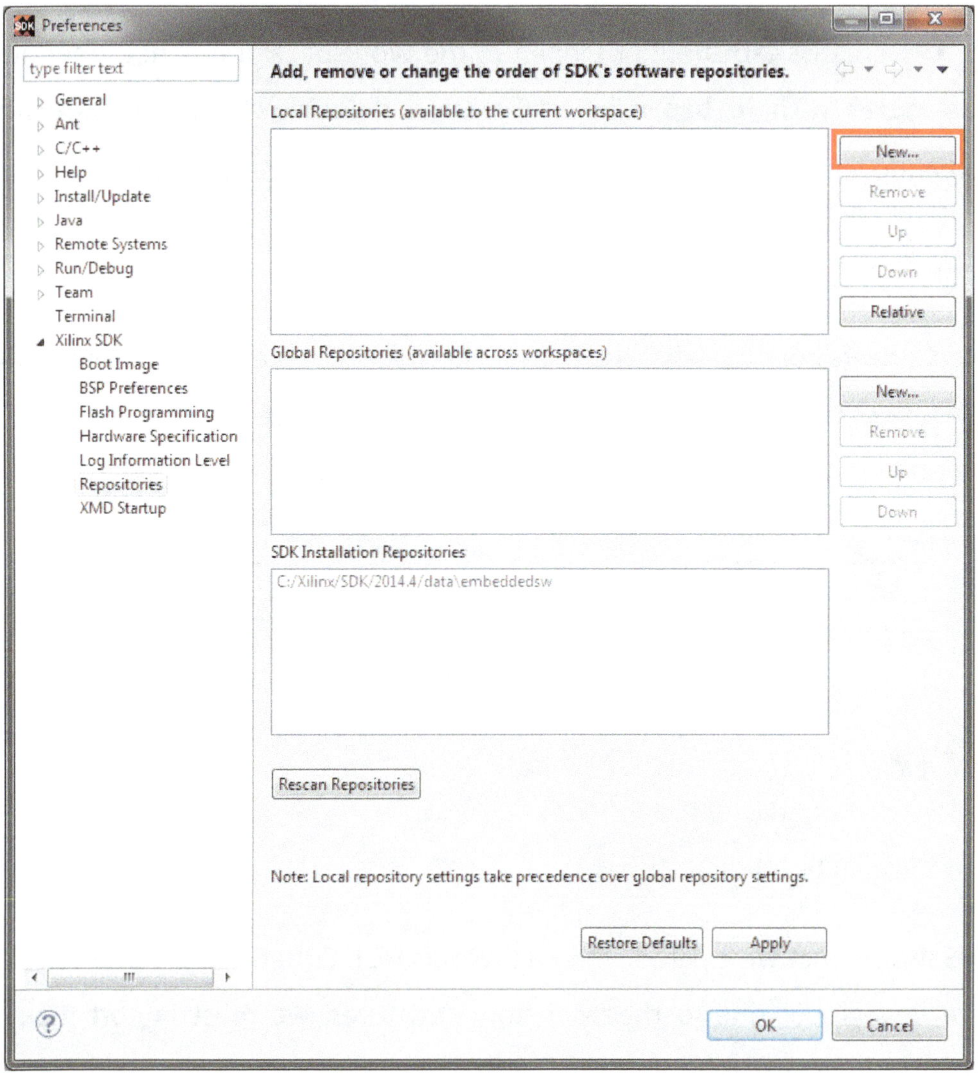

Figure 5.31: SDK Repository Peripherals window

(d) Add the LED Controller drivers by browsing to the directory:

C:\Zynq_Book\ip_repo\led_controller_1.0 and clicking ***OK***.

(e) Click ***New***.

Add the NCO drivers by browsing to the directory:

C:\Zynq_Book\ip_repo\xilinx_com_hls_nco_1_0

and clicking ***OK***.

Upon closing the preferences window, SDK will automatically scan the repository and rebuild the project to include the driver files.

We must now check that the newly imported drivers have been assigned to their corresponding peripherals.

(f) The **system.mss** tab should be open in the Workspace. If it is not, open it by expanding **adventures_with_ip_bsp** in *Project Explorer* and double-clicking on ***system.mss.***

(g) At the top left of the **system.mss** tab, click ***Modify this BSP's Settings***.

The *Board Support Package Settings* window will open.

(h) Select drivers from the left-hand menu and check that the **led_controller** driver is assigned to the **led_controller_0** component and the **nco_top** driver is assigned to the **nco_0** component, as highlighted in Figure 5.32.

led_controller_0	led_controller	led_controller	1.00.a
lms_pcore_0	lms_pcore	generic	1.00.a
nco_0	nco	nco_top	1.00.a

Figure 5.32: Driver assignment

Click ***OK***.

The project will now rebuild.

The LMS IP core that we created with MathWorks HDL Coder and the audio codec IP also have software drivers, but due to their directory structures, we must import their drivers to the workspace rather than use a repository.

(i) In the *Project Explorer* panel, expand **adventures_with_ip**, right-click on **src** and select ***Import***.
In the *Import* window, expand **General** and double-click on ***File System***.
Click ***Browse*** in the top right corner, and navigate to
C:\Zynq_Book\hdl_coder_lms\hdl_prj\ipcore\lms_pcore_v1_00_a\include.
Click ***OK***, to import the LMS IP driver.
In the right-hand panel, select ***lms_pcore_addr.h*** and click ***Finish.***
Note: This directory will only be available if you have completed **Exercise 4B** of **Tutorial 4**.
If you have not completed this exercise, you can obtain ***lms_pcore_addr.h*** from the Zedboard directory:
C:\Zynq_Book\sources\zedboard\adventures_with_ip_integrator\drivers

or the Zybo directory:

C:\Zynq_Book\sources\zybo\adventures_with_ip_integrator\drivers

(j) The **audio.h** header file should be imported using the correct directory depending on your chosen Zynq development board.

The *Zedboard* directory:

C:\Zynq_Book\sources\zedboard\adventures_with_ip_integrator\drivers.

The *Zybo* directory:

C:\Zynq_Book\sources\zybo\adventures_with_ip_integrator\drivers.

(k) With all the driver files for the IP imported, we can import the source files for our application.

Zed Follow the same procedure as in **Step (i)** to import the following files from the **C:\Zynq_Book\sources\zedboard\adventures_with_ip_integrator\software** directory:

- **adventures_with_ip.h**
- **adventures_with_ip.c**
- **audio.c**
- **ip_functions.c**

Zybo Follow the same procedure as in **Step (i)** to import the following files from the **C:\Zynq_Book\sources\zybo\adventures_with_ip_integrator\software** directory:

- **adventures_with_ip.h**
- **adventures_with_ip.c**
- **audio.c**
- **ip_functions.c**

Resume The source files will be imported and the application should build.

(l) Open the header file **adventures_with_ip.h** by double-clicking on it in *Project Explorer*.

This is the main header file for the software application. At the top of the file you should see a list of included header files, which define a variety of functions which are used in the software application.

Further down the file you should see the inclusion of the custom IP header files which we imported earlier:

```
/* ------------------------------------------------------------------------------ *
 *                          Custom IP Header Files                                  *
 * ------------------------------------------------------------------------------ */
#include "audio.h"
#include "lms_pcore_addr.h"
#include "xnco.h"
```

As an example of one of the header files that was created during the IP creation process, we will open the header for the LMS IP core.

(m) In the *Outline* tab on the right hand side of the SDK window, double click on **lms_pcore_addr.h**.

In the LMS header file, you should see the following definitions:

```
#define  IPCore_Reset_lms_pcore    0x0  //write 0x1 to bit 0 to reset IP core
#define  IPCore_Enable_lms_pcore   0x4  //enabled (by default) when bit 0 is 0x1
#define  IPCore_Strobe_lms_pcore   0x8  //write 1 to bit 0 after write all input data
#define  IPCore_Ready_lms_pcore 0xC  //wait until bit 0 is 1 before read output data
#define  x_k__Data_lms_pcore       0x100  //data register for port x(k)
#define  d_k__Data_lms_pcore       0x104  //data register for port d(k)
#define  e_k__Data_lms_pcore       0x108  //data register for port e(k)
```

These define the **memory-mapped address offsets** of the various signals of the LMS peripheral. Data can be transferred between the peripheral in the PL and the software in the PS by writing to, or reading from the these offset addresses. The actual address that would be used to access these signals would be **BASE ADDRESS + OFFSET**.

Each IP peripheral that we added to our block design in IP Integrator is automatically assigned a base address in memory. These addresses can be determined from a *Xilinx parameters* C header file which is automatically created when exporting an IP Integrator design that contains a Zynq Processing System. The header file is called **xparameters.h**.

We shall now explore the *Xilinx parameters* header file.

(n) Switch back to the **adventures_with_ip.h** tab in the *Editor* window.

xparameters.h is included in this main header file, and is therefore accessible from the *Outline* tab.

(o) Open **xparameters.h** by double-clicking on it in the *Outline* tab.

Here you should see a list of memory *base address* definitions, along with a number of other parameters.

As we were previously looking at the LMS header file, we will look at the definition of the base address for the LMS peripheral.

(p) Scroll down the file until you see the following lines:

```
/* Definitions for peripheral LMS_PCORE_0 */
#define XPAR_LMS_PCORE_0_BASEADDR 0x43C00000
#define XPAR_LMS_PCORE_0_HIGHADDR 0x43C0FFFF
```

Here we see the definitions of both the base and high addresses in memory for the LMS peripheral. This will vary depending on the value the *Base Address* was set to in Exercise 5B, Step (p). The address range was set to a value of **64 Kilo-Bytes**. The difference between the high address and the base address is 0xFFFF, the LMS peripheral has an addressable range of **65536 bits**, or **64 Kilo-Bytes**.

Referring back to the **memory-mapped address offsets** for the LMS block in Step (m), if we, for example, wanted to write data to the input port **x(k)**, we would do this by writing the desired value to the BASE ADDRESS + OFFSET, which in this case would be:

XPAR_LMS_PCORE_0_BASEADDR + x_k__Data_lms_pcore = 0x43C10000 + 0x100

Giving a unique address of **0x43C10100**.

We will now take a look at the main software application file.

(q) Open the source file **adventures_with_ip.c** by double-clicking on it in *Project Explorer*.

This file contains the main function, and another function which implements an interactive menu that allows the user to control the system using keyboard commands via the terminal.

Take a moment to look over the file and note the function calls which are made.

In the `main()` function, the first set of functions are called to setup and configure the audio codec. These functions are defined in **audio.c**, which we will look at next.

(r) Open **audio.c**.

Here we have the functions which are called to initialise the audio codec and the required I^2C interface in the Zynq PS.

We don't want to go into great detail about the functionality contained here, but in basic terms the purpose of these functions is to configure the audio codec by writing to the codec's control registers.
Each control register has a unique address which can be accessed via the I^2C serial interface.
The control register addresses are defined in the **audio.h** header file.

(s) Open **audio.h**.

This file contains a number of definitions relating to the audio codec and the I^2C interface, as well as some prototype function definitions.

You should see an enumerated type which lists all of the audio codec's control register addresses, which were mentioned in the previous step.

More information on the audio codecs for both the Zedboard and Zybo can be found in the following data sheets respectively:
http://www.analog.com/static/imported-files/data_sheets/ADAU1761.pdf

http://www.analog.com/media/en/technical-documentation/data-sheets/SSM2603.pdf

Next we will have a look at the functions which control the custom IP peripherals in the PL.

(t) Open **ip_functions.c**.

This file contains the functions which control the IP peripherals, as well as some functions to initialise drivers for the GPIO and NCO.

The three functions of interest are:

- **`audio_stream()`** — Implements stereo audio loopback between the input and output ports of the audio codec. Left and right audio samples are read in from the audio controller peripheral's I^2S receive register and are then written back out to the controller's I^2S transmit register.
- **`tonal_noise()`** — This function builds upon the audio loopback in **audio_stream()**. A step size value is input via the slide switches on the board. The corresponding value is then output to the LEDs on the board by writing to the memory-mapped register of the LED controller peripheral. The step size value is also output to the NCO peripheral using the **`XNco_Set_step_size_V()`** function defined by the NCO driver file. A sinusoidal sample created by the NCO peripheral is the read in by the **`XNco_Get_sine_sample_V()`** NCO driver function and, as in the previous audio streaming function, left and right audio samples are received from the audio codec. The sinusoidal noise component is then added to the left and right audio samples before being written to the audio controller for output to the codec.
- **`lms_filter()`** — This function combines the functionality of the NCO and the LMS peripherals to create a system which adds tonal noise to an audio signal, before using an LMS adaptive filter for noise cancellation to remove the added noise. As in the **`tonal_noise()`** function, sinusoidal samples are generated from the NCO peripheral and added to the left and right audio samples from the audio controller. The sinusoidal sample is then input to the LMS as the input sample **`x(k)`** and the sample with added tonal noise is input as the desired signal **`d(k)`**. The resulting output of the LMS peripheral is only read if the user presses any of the push buttons on the board, otherwise the corrupted audio sample is retained. This allows the user to verify that the LMS filter peripheral is removing the noise.

Now that we have had a look at the functions and definitions contained in the various source and header files, we can move on to actually implementing the system on the Zynq development board.

To begin, we will program the Zynq PL with the bitstream that we generated in the previous exercise.

Note: At this stage ensure that the Zynq development board is powered on and both the **PROG** and **UART** USB ports are connected to your host computer. The Zybo has a single USB port for both **PROG** and **UART** connections and the Zedboard has two USB ports, one for **PROG** and another for **UART**.
You should also ensure that the board is configured to boot from **JTAG**.

(u) Select ***Xilinx Tools*** > ***Program FPGA*** from the *Menu Bar*. The Program FPGA window should be configured as in Figure 5.33.

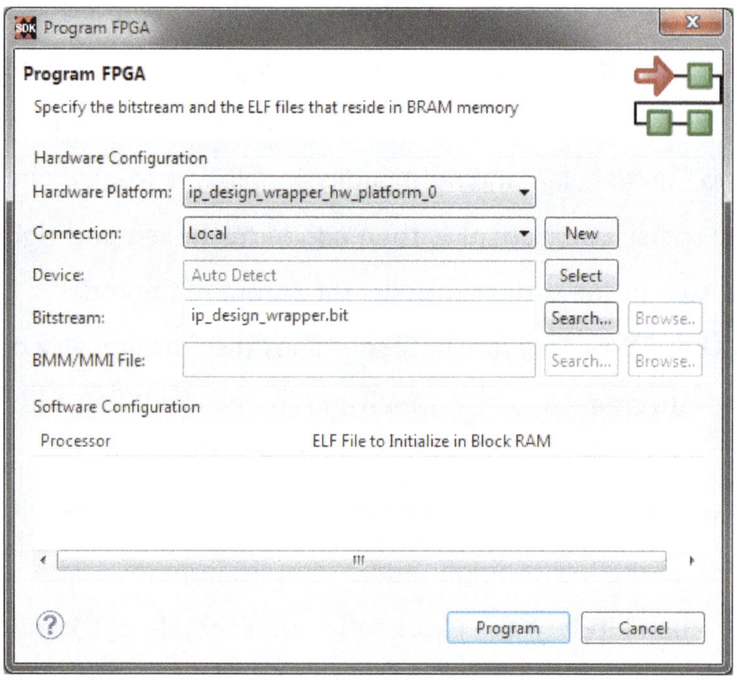

Figure 5.33: Program FPGA window

Click ***Program***.
The Zynq PL on the board will be configured with the bitstream and the **DONE** LED should illuminate.

At this stage we must invoke PuTTY — the terminal program which you should have downloaded at the beginning of this tutorial.

(v) At the location which you downloaded PuTTY, double-click ***PuTTY.exe***. As you downloaded the executable file, Windows may present a security warning. Accept the warning by clicking ***Run***.

(w) *PuTTY Configuration* should open, as shown in Figure 5.34.

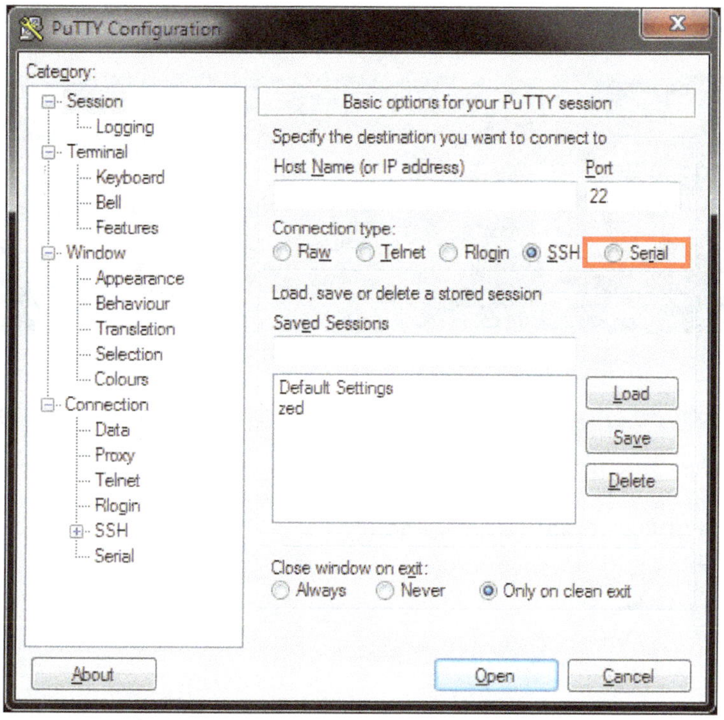

Figure 5.34: PuTTY

(x) Select ***Serial*** as *Connection type* (highlighted in Figure 5.34) and configure the settings as specified in Figure 5.35.

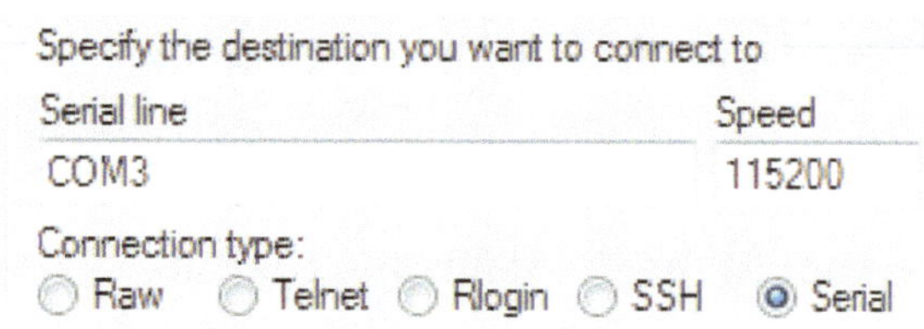

Figure 5.35: PuTTY configuration

NOTE: The value of the *Serial line* entry will vary depending on which the USB UART cable is connected to.

In order to determine this value on a Windows system, open the Device Manager and identify the COM port which may be named 'USB Serial Port'.

(y) Click ***Open***, to open the terminal connection. The PuTTY terminal window will open.

With the terminal connection open, the final step is the run the software on the Zynq PS.

(z) Right-click on ***adventures_with_ip*** in Project explorer and select ***Run As > Launch on Hardware (GDB)***.

Zed In the PuTTY terminal you should see the following output:

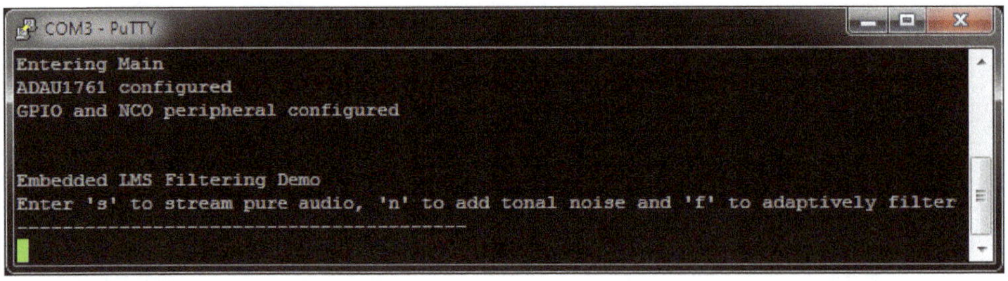

Note: At this point you should attach an audio patch cable between the PC speaker output and the board's *LINE IN* input. Also, connect headphones to the board's *LINE OUT* input. These connections are highlighted in Figure 5.36.

Figure 5.36: ZedBoard Audio Jacks

Zybo In the PuTTY terminal you should see the following output:

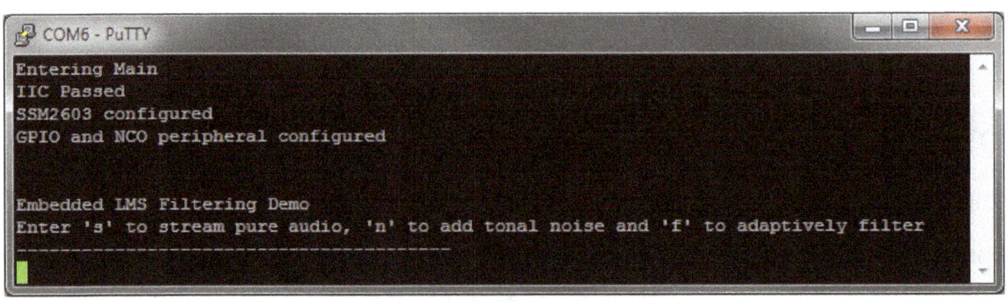

Note: At this point you should attach an audio patch cable between the PC speaker output and the board's *LINE IN* input. Also, connect headphones to the board's *HPH OUT* input. These connections are highlighted in Figure 5.37.

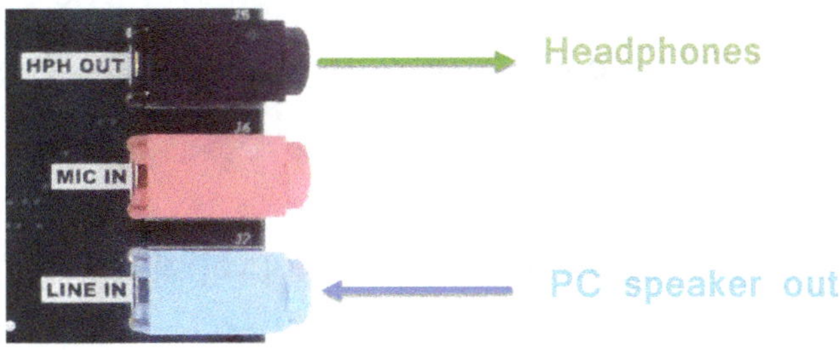

Figure 5.37: Zybo Audio Jacks

Resume (aa) Open the audio file

C:\Zynq_Book\sources\input\original_speech.wav

in an audio player, and begin playback.

Note: It may be useful to turn on the repeat setting in the audio player for continuous playback.

(ab) Set all switches on the Zynq development board to the 'off' position.

(ac) In the **PuTTY** terminal window, press the ***'s'*** key on your keyboard.

This will prompt the software application to enter the **`audio_stream()`** function which we looked at earlier.

You should be able to hear audio of speech via the headphone connection.

(ad) Press the ***'q'*** key on the keyboard to return to the **menu**.

(ae) Press the ***'n'*** key on the keyboard. This will prompt the application to enter the **`tonal_noise()`** function.

Initially you should hear the same audio signal.

You should note that currently there is no step size being input to the NCO.

Push slide switch **SW0** into the on position. You should now be able to hear a sinusoidal tone which has been added to the audio signal. **LED 0** should also be lit.

Experiment with different step size values by varying the on/off values of slide switches **SW1** and **SW2**. This will vary the frequency of the tonal noise. Note the updates in Putty.

(af) Press the ***'q'*** key on the keyboard to return to the **menu**.

(ag) Press the ***'f'*** key on the keyboard. This will prompt the application to enter the **`lms_filter()`** function. The basic functionality here is the same as in the previous NCO function, and you can add tonal noise to the audio signal using the slide switches.

With tonal noise being added to the audio signal, press and hold any of the push buttons on the board. The sinusoidal tone will be adaptively filtered by the LMS, and the tonal noise removed.

This concludes this exercise on the creation of an audio application in the SDK. You should now be familiar with:

- The automatically generated **xparameters.h** header file, and its contents.
- Identifying memory-mapped base addresses and offsets for communication between software running on the Zynq PS and peripherals in the PL.
- The procedure of configuring the ZedBoard's ADAU1761 and Zybo's SSM2603 audio codec via the control register addresses.
- Receiving and sending audio samples to/from the audio codec via an audio controller block in the PL.
- The process of communicating with custom peripherals in the PL via generated software drivers.

CPSIA information can be obtained
at www.ICGtesting.com
Printed in the USA
BVHW01s0257210818
524965BV00006B/35/P

9 780992 978730